SCIENCE, TECHNOLOGY AND SOCIETY

Photocopiable Resource Pack of Related Interest

Essential Science Activities for Key Stage 4
Keith Bishop • Bill Scott • David Maddocks

SCIENCE, TECHNOLOGY AND SOCIETY

David Andrews
Principal Tutor (Curriculum)
Havering Sixth Form College

Stanley Thornes (Publishers) Ltd

Text © D. Andrews.

Original line illustrations © ST(P) Ltd 1992.

The copyright holders authorise ONLY purchasers of this pack to make photocopies or stencil duplicates of the Units for their own or their classes' immediate use within the teaching context.

Photocopies or stencilled duplicates must be used immediately and not stored for further use or re-use.

No other rights are granted without permission in writing from the publisher or under licence from the Copyright Licensing Agency Limited. Further details of such licences (for reprographic reproduction) may be obtained from the Copyright Licensing Agency Limited, 90 Tottenham Court Road, London W1P 9HE.

Copy by any other means or for any other purpose is strictly prohibited without the prior written consent of the copyright holders.

Applications for such permission should be addressed to the publishers: Stanley Thornes (Publishers) Ltd, Old Station Drive, Leckhampton, CHELTENHAM GL53 0DN, England.

First Published in 1992 by:
Stanley Thornes (Publishers) Ltd
Old Station Drive
Leckhampton
CHELTENHAM GL53 0DN
England

British Library Cataloguing-in-Publication Data

Andrews, David
   Science, technology and society.
   I. Title
   501

ISBN 0-7487-1293-3

Artist: Mike Parsons, Barking Dog Art
Typeset in Plantin by Tech-Set, Gateshead, Tyne & Wear.
Printed and bound by Dotesios Ltd, Trowbridge, Wilts.

Contents

Introduction — x

1 SCIENCE – THE KEY TO THE UNIVERSE?

1.1 What is science? — 1
1.2 Religion and evolution — 3

2 LAUNCHING TECHNOLOGY

2.1 The space race: competition or co-operation? — 6
2.2 The *Challenger* Shuttle disaster — 8
2.3 Future space exploration: a mission to Mars? — 10
2.4 Industry and innovation — 12
2.5 Industry and the environment: bauxite mining — 14

3 INFORMATION TECHNOLOGY

3.1 Trends in information technology — 17
3.2 From valves to transistors to microchips — 19
3.3 Computer applications — 21
3.4 Artificial Intelligence — 22
3.5 The Data Protection Act: who is holding information about you? — 24
3.6 Could a computer do your job? — 26
3.7 Marketing innovations: satellite television — 29

4 WAR, TECHNOLOGY AND SOCIETY

4.1 Warfare and technological development — 31
4.2 The atomic bomb: from Einstein to Hiroshima — 33
4.3 Atomic bomb decisions — 35
4.4 The medical effects of nuclear war — 36
4.5 Controlling the arms race — 38
4.6 Chemical weapons — 40

5 THE REPRODUCTION REVOLUTION

5.1 Natural selection at work — 42
5.2 Basic genetics — 44
5.3 Harmful genes 1: sickle cell anaemia — 47
5.4 Harmful genes 2: cystic fibrosis — 49
5.5 Amniocentesis and Down's Syndrome — 51
5.6 Images of disability — 54
5.7 Mass screening for inherited disorders — 56
5.8 Screening embryos and fetuses — 58
5.9 Genetic counselling — 60

6	IMPROVING THE SPECIES?	
6.1	Test tube babies	62
6.2	Embryo research choices: Germany and the UK	63
6.3	Freezing embryos	66
6.4	Surrogate motherhood: rent a womb?	68
6.5	Genetic engineering	70
6.6	Gene therapy	72
6.7	Who owns the human genome?	73

7	FEED THE WORLD	
7.1	Good for you?	74
7.2	Food additives	75
7.3	Food advertising and labelling	76
7.4	Mycoprotein	78
7.5	Intensive agriculture and the environment	80
7.6	The Green Revolution	82

8	MEDICINE, HEALTH AND SOCIETY	
8.1	Health care: curative and preventative approaches	84
8.2	Appropriate health care: barefoot doctors	86
8.3	AIDS/HIV campaigns	88
8.4	Health and social class	90
8.5	Attitudes to abortion	92
8.6	Medical use of fetal tissue	94

9	WORLD POPULATION – PROBLEMS AND SOLUTIONS	
9.1	Population growth: the demographic transition	97
9.2	Population pyramids	100
9.3	Contraceptive choice	102
9.4	Population control policy	105

10	A PILL FOR EVERY ILL?	
10.1	Drug development: the sulphonamide story	107
10.2	Testing drugs on humans and animals	108
10.3	Tackling the drug problem	110
10.4	Drugs in the developing world: inappropriate technology?	112

11 ENERGY – RESOURCES AND CONSUMPTION

11.1	Energy basics	114
11.2	Fossil fuels	116
11.3	Supplying electricity	118
11.4	International comparisons of energy consumption	120
11.5	Patterns of energy use in the UK	121
11.6	Trends in energy consumption in the UK	122
11.7	Energy, transport and pollution	124
11.8	Save it!	126
11.9	Global warming	128

12 RENEWABLE ENERGY RESOURCES

12.1	Solar energy	130
12.2	Biomass energy	133
12.3	Hot rock power: geothermal energy	135
12.4	Tidal power: a feasibility study	136
12.5	Hydropower: large-scale and small-scale	139
12.6	The potential of wind power	140
12.7	Wind power in the developing world	142
12.8	Wind and wave power under attack	144
12.9	The future of renewable energy technologies	147

13 NUCLEAR POWER

13.1	What is radioactivity?	149
13.2	How do nuclear power stations work?	151
13.3	Reprocessing and breeder reactors	153
13.4	What can we do with radioactive waste?	154
13.5	The Chernobyl accident	157
13.6	Nuclear power and leukaemia: problems of extrapolation	158
13.7	Hot and cold fusion power	160

Teacher's notes   162

Sources of further information   181

Acknowledgements

The author and publisher wish to thank the following for permission to reproduce text and illustrations. Every effort has been made to trace the copyright and we apologise for any omissions.

p. 5, *The Observer*. p. 7, Primo Levi, *The Mirror Maker*, translated from the Italian by Raymond Rosenthal, published by Methuen, London © 1986 Editrice *La Stampa* s.p.a., Torino. Translation © 1989 by Schocken Books Inc. p. 12, Fig. 2.4.1 The Motoring Picture Library, The National Motor Museum at Beaulieu. p. 13, © Times Newspapers Ltd. 1989. (Photocopying allowed for classroom use only.) p. 19, Fig. 3.2.1 Philips Research Laboratories, Redhill, England. p. 25, Data Protection Registrar. p. 32, *New Scientist*. p. 34, Scherz Verlag. pp. 36, 37, Fig. 4.4.1, Tables from *Report on the social and medical consequences of Japanese atomic bombs*, Hutchinson Ltd, 1979. p. 38, Campaign for Nuclear Disarmament. p. 46, Fig. 5.2.4 Science Source/Science Photo Library. p. 47, Fig. 5.3.1 Omikron/Science Photo Library. p. 50, *The Independent on Sunday*. p. 52, Fig. 5.5.2 Philip Harris Education, Fig. 5.5.3 Dept. of Clinical Cytogenetics, Addenbrookes Hospital, Cambridge/Science Photo Library. p. 54, Fig. 5.6.1 Down's Syndrome Association. p. 55, *The Guardian*, Fig. 5.6.2 Robert Sone. p. 65, *Nature* vol. 348 © 1990 Macmillan Magazines Ltd. p. 71, Bayer AG. p. 76, Fig. 7.3.1 Sainsbury's de Luxe Muesli, Crosse & Blackwell, Fox's Biscuits. p. 79, Fig. 7.4.1 Elizabeth Sewell at Marlow Foods. p. 81, *New Scientist*. p. 82, *Small is Beautiful*, E. Schumacher, Sphere Books Ltd. p. 87, Reprinted with the permission of the Peters Fraser & Dunlop Group. p. 96, Sarah Boseley, *The Guardian*. p. 103, Fig. 9.3.1 Organon Laboratories Ltd. p. 107, *research*, the scientific magazine of Bayer AG. p. 113, *The Observer*. p. 117, Quotation from *A Blueprint for Survival* courtesy of Rogers, Coleridge & White Ltd. Quotation of Prof. Peter Odell courtesy of *The Geographical Journal*. p. 119, CEGB Annual Reports and Accounts 1987/88. p. 120, Table 11.4.1 reproduced from *BP Statistical Review of World Energy* 1986. pp. 131-4, 137-8, 148, Excerpts and diagrams reprinted by permission of the UK Department of Energy. p. 143, Fig. 12.7.1 Intermediate Technology Group, Rugby. pp. 145-6, *New Scientist*. pp. 154-6, Figs. 13.4.1, 13.4.2 United Kingdom Nirex Ltd. p. 156, Fig. 13.4.4 AEA Technology. p. 157, *Daily Mail, Daily Mirror*. p. 158, Fig. 13.6.1 from *Report on the social and medical consequences of Japanese atomic bombs*, Hutchinson Ltd, 1979. pp. 160-1, The JET Joint Undertaking.

Introduction

Scientific knowledge cannot be pursued without reference to the social, political, economic and ethical dimensions of the discoveries made. These elements have been incorporated into the National Curriculum and also form an important part of the proposed common core for 16–19 education.

Most teachers would agree that science must be seen as relevant to the students' own experience and concerns. Not only potential scientists, but also those who will be on the receiving end of scientific discoveries, should be given the opportunity to develop informed opinions on scientific advances and the issues that surround them. *Science, Technology and Society* seeks to aid this development through encouraging active involvement by students, including discussion, role-play, simulations and data analysis.

Each unit includes one or more activities, most of which are based on actual people and events, designed to investigate real-life problems and solutions. Many units summarise key technological advances and recent scientific trends. To keep the material topical in the future, teachers can supplement the units with information on the latest developments. For example, changes in government policy and advances in scientific knowledge can have a considerable impact on technological developments such as nuclear fusion, gene therapy and pollution control.

The 'Teacher's Notes' provide answers to questions, where relevant, guidance on discussion points, and suggestions for further reading. 'Sources of Further Information' is a list of useful addresses.

I should like to thank my colleagues for all their help in preparing these materials, Katherine Pate for her careful editing and proof-reading and Adrian Wheaton for his advice.

David Andrews

1 SCIENCE – THE KEY TO THE UNIVERSE?

UNIT 1.1 What is science?

'Science is what is done in laboratories by people in white coats.' This simple view of science is a rather narrow one. Another more useful way of looking at science is to see it as a systematic way of thinking and problem solving that all of us use, not only scientists.

If you look at the colour of the clouds in the morning to help you make up your mind whether or not to take an umbrella, you are thinking in a scientific way. The main steps in this process are as follows:

1. On previous mornings you have made observations about the type of clouds in the sky and during the day you have registered whether or not it rains. Accurate observations and measurements form the basis of the scientific approach.

2. On the basis of these observations you develop your own 'theory' about rain clouds. Like a scientist, you express your theory as a generalisation that explains the link between the colour of the clouds and the risk of rainfall: dark clouds are usually linked with rain.

3. You use your theory to make predictions about the future and to help you make decisions. You decide to take your umbrella if there are dark clouds in the sky.

4. A thoughtful scientist will always be on the lookout for observations that might question the truth of a particular theory. These anomalous observations can lead to the rejection of an old theory and trigger the search for a new one. Your rain cloud theory will be undermined if a dark, cloudy sky does not produce rain. You will modify or completely discard your theory, just as scientists do, if new evidence comes to light.

SCIENTIFIC PARADIGMS

A **paradigm** is a framework of evidence and theories that scientists use as a springboard for their investigations of the world. They will try to slot all their new observations into the accepted paradigm. But paradigms are only temporary frameworks. Evidence that contradicts the paradigm may start to accumulate. Eventually the weight of evidence against the paradigm will bring it crashing down. A scientific revolution occurs and a new paradigm must be installed in place of the old one. This happened when the Earth-centred view of the universe was rejected by astronomers in the seventeenth century. It is possible for two competing paradigms to exist at the same time with scientists debating fiercely about which is the right one. It may take many years to arrive at a paradigm that is agreed by all scientists in a particular field. At present there are two paradigms concerning the origin of the universe, that physicists disagree on.

ACTIVITY

1. Compare a belief in horoscopes with the process of thinking outlined on the left. Does it count as scientific?

2. A scientific experiment is a way of making precise observations of events while keeping a tight control over all the conditions that might affect the subject you are studying. With this in mind, devise and carry out experiments to test the truth of the phenomenon known as telepathy (the transmission of images or thoughts from one person to another without visual or verbal communication).

3. Physicists can make accurate predictions about the material world. For example they could give the exact velocity of an object dropped from a height as it hits the ground. Meteorologists gather together a vast array of data to make predictions about the weather. Yet sometimes their forecasts turn out to be wrong. Does this mean that physics is a science but meteorology is not?

4. Scientists disagree over many important issues in science. For example, there are wildly differing estimates of the seriousness of the health risks of the radioactivity released by the accident at the Chernobyl nuclear power station in 1986. Will science ever progress to the stage where all scientists can reach an agreement on questions like this?

COPERNICUS AND THE REVOLUTION IN ASTRONOMY

Humans have always sought to understand the movements of the visible 'heavenly bodies' as, night after night, they have watched them progress slowly across the sky. An early paradigm established by Ptolemy, the Greek astronomer, in AD 120, placed the Earth at the centre of the universe and was accepted for over 1500 years. The Sun, Moon, planets and stars were thought to be embedded in their own solid but transparent 'celestial spheres', which revolved around the Earth. This paradigm made sense of the fixed arcs the heavenly bodies trace as they cross the sky. Unfortunately, closer observation of the paths of some of the planets did not fit the paradigm. These planets seemed to go into reverse periodically before making further progress across the night sky. To account for this 'retrograde motion' Ptolemy added extra smaller celestial spheres attached to the larger ones (Figure 1.1.1). This complicated system was accepted in the absence of a suitable alternative but one medieval astronomer voiced the dissatisfaction of many when he said, 'If God had consulted me before the Creation, I would have recommended something simpler.'

In the sixteenth century a simpler paradigm was suggested by the Polish mathematician Copernicus, which put the Sun at the centre of the universe with all the planets, including the Earth, revolving around it.

The Copernican paradigm caused a turbulent scientific revolution, as its supporters faced concerted opposition from the Catholic Church. The church leaders at the time believed that the Earth-centred universe was a God-given truth. Anyone denying it was denounced as a heretic. The Italian astronomer Galileo collected convincing evidence from his observations of the planets throughout his life, but the church banned his books and he was forbidden to teach.

COMPETING PARADIGMS IN COSMOLOGY

The origins and history of the universe are investigated by a branch of science called cosmology. For the past 40 years two alternative paradigms have been subject to heated debate – the Big Bang and the Steady State. In the late 1940s the Steady State theory, proposed by Fred Hoyle, assumed that the universe has changed very little throughout its history and that new matter is continually being created. During the 1970s the Big Bang theory gained support from astronomical observations that suggested that the universe originated from a massive explosion and that every galaxy is now rushing away from every other galaxy. However, there is still a strong dissenting group who believe that the evidence supports the idea of a steady state universe.

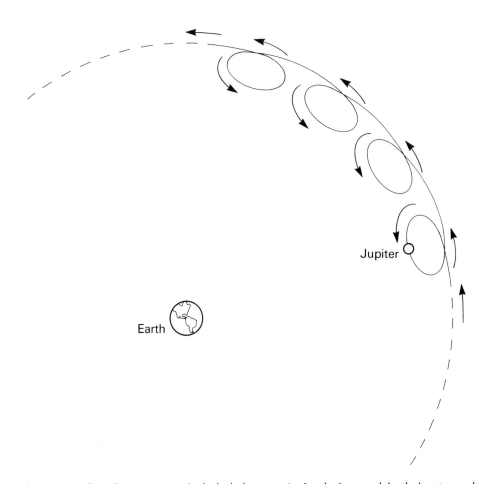

Figure 1.1.1   Ptolemy's paradigm for astronomy included planetary 'epicycles' to explain their retrograde motion.

1 SCIENCE – THE KEY TO THE UNIVERSE?

UNIT 1.2 Religion and evolution

There are over 200 million different species of plants and animals on our planet. Where have they all come from? Many cultures around the world answer this question by a belief in a supernatural being who created the Earth and all living things at a particular point in time. In Europe, until the nineteenth century, the religious belief in God, the Creator, was largely unquestioned. The account given in Genesis, the first book of the Old Testament of the Bible, was taken as literal truth.

At the beginning of the nineteenth century scientists had begun to accumulate evidence of an alternative explanation – the theory of evolution. It suggested that organisms change their form over many generations, one species giving rise to a number of different species, so that simpler life forms develop into more complex ones over millions of years. The origin of life was explained by the evolution of simple life forms from non-living material. Although scientists readily adopted this new theory and discarded the biblical version, it gave rise to many bitter controversies and even today, for some people, the issue remains unsettled. For them the religious and scientific explanations are incompatible. But for others there is no problem in reconciling their beliefs in evolution *and* religion. The comparison of the scientific and religious approaches in this unit should help you to gain a clearer idea of your own position on this question.

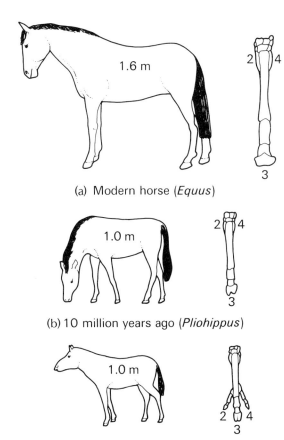

(a) Modern horse (*Equus*)

(b) 10 million years ago (*Pliohippus*)

(c) 30 million years ago (*Merychippus*)

Figure 1.2.1 — continued.

ACTIVITY

Creationism

1 A belief known as Creationism proposes that the universe and all plant and animal life, including human beings, were created at the same time by God, and that since that time, life forms have remained unchanged.
 - What is your attitude to Creationism?
 - What does a Creationist use to support his or her view?

2 In the eighteenth century the discovery of fossils of extinct organisms, such as dinosaurs, and the existence of fossil series such as the horse's leg (see Figure 1.2.1) started to raise problems for the Creationist view.
Suggest why this was seen as a problem.

3 Philip Henry Gosse was an eminent Victorian zoologist and a committed Christian who struggled to reconcile the fossil evidence with his religious beliefs. His explanation of the fossil record was that it had been placed into the rocks by God in order to test the strength of people's faith in Him. Others were sceptical of the idea that God had deliberately formed evidence that might raise doubts about His existence.
What is your view of this argument?

(d) 40 million years ago (*Mesohippus*)

(e) 60 million years ago (*Hyracotherium*)

Figure 1.2.1 The fossils of the ancestors of the horse form a sequence that suggests there has been gradual evolutionary change.

ACTIVITY

1 Some of the main religious objections to evolution and the counter arguments are given in Table 1.2.1. Discuss the merits of each point of view.

2 In the USA there has been pressure for the Creationist view to be given equal time and status with the teaching of evolution in biology lessons. This has been strongly opposed by science teachers.
What is your view on this controversy?

Table 1.2.1

Creation	Evolution
Humans are totally different to other animal species. They could not have evolved from apes and must have been created independently.	There is evidence from a variety of sources to show that humans have close evolutionary links with apes. The fossil record shows the gradual evolution of the skull and skeleton from ape-like to a more human form. Biochemical analysis shows that 98 per cent of the genes in humans and gorillas are identical.
The intricacy and perfect form of structures such as the mammalian eye could not have arisen through random blind forces. They show the work of intentional design by a Creator.	The appearance of conscious design is a result of natural selection. Its mechanism of the 'survival of the fittest' ensures that random advantageous variations will accumulate gradually to produce refinements of increasing complexity. (See Unit 5.1.)

RICHARD HARRIES

A myth that will not die

OBSCURANTIST religion 'opposes every new scientific discovery' — that is the myth, seemingly impossible to eradicate. Nowhere has it been etched deeper than in the view that the Church opposed Darwin's theory of evolution, a hostility dramatically enacted in the encounter between Huxley and 'Soapy Sam' Wilberforce at the British Association meeting in Oxford in 1860. The legend, endlessly repeated in books and films, is well given 40 years after the event in the October 1898 edition of *Macmillans Magazine* in an article entitled 'A Grandmother's Tale':

> The Bishop rose, and in a light scoffing tone, florid and fluent, he assured there was nothing in the idea of evolution; rock-pigeons were what rock-pigeons had always been. Then turning to his antagonist with smiling insolence, he begged to know, was it through his grandfather or his grandmother that he claimed his descent from a monkey? On this, Mr Huxley slowly and deliberately arose. A slight tall figure, stern and pale, very quiet and very grave, he stood before us, and spoke those tremendous words — words which no one seems sure of now, nor I think, could remember just after they were spoken, for their meaning took away our breath, though it left us in no doubt as to what it was. He was not ashamed to have a monkey for his ancestor; but he would be ashamed to be connected with a man who used great gifts to obscure the truth. No one doubted his meaning and the effect was tremendous. One lady fainted and had to be carried out: I for one jumped out of my seat.

The truth of the matter is rather different. Darwin's theory was indeed fiercely opposed at the beginning; but by scientists on scientific grounds. Some religious people opposed it on religious grounds, but others did not.

The sermon at the British Association meeting that year was preached by Frederick Temple, then Headmaster of Rugby and later Archbishop of Canterbury. He argued strongly that God works through Laws of Nature; it was not impious to think of God working through impersonal forces in a long process of evolution. 'Men were thought impious who attempted to represent . . . the existence of animal life as the work of natural causes, and not the direct handiwork of God himself. And yet . . . there seems no more reason why the solar system should not have been brought into its present form by the slow working of natural causes than the surface of the earth, about whose gradual formation most students are now agreed.'

A good number of distinguished Christians came out early in favour of evolution and the American scientist, Asa Gray, a close friend of Darwin, said that in his opinion there had been no undue reluctance on the part of Christians in accepting evolution.

Darwin himself was a humble, lovely man, who was nearly ordained earlier in his life. When he wrote 'The Origin of Species' he still believed in God and, although his agnosticism grew over the years, he never ruled out the possibility of the existence of a creator. He regarded his theory of evolution as quite compatible with religious belief. What eroded his faith was historical criticism of the New Testament and his detailed attention to the actual workings of nature, which seemed, by human standards, to contain so much pain and cruelty.

© *The Observer*, 4 September 1988

ACTIVITY

The great debate

The passage above is from an article by Richard Harries, then Bishop of Oxford, first published in 1988. It refers to a historical confrontation in a public debate on religion and evolution, between two well-known personalities – Bishop Wilberforce, nicknamed 'Soapy Sam', and Thomas Huxley, one of Darwin's staunchest allies. It is one of many confrontations between religious believers and evolutionists that have occurred since Darwin's time.

1 Find out what is meant by 'obscurantist religion'.

2 What is Harries' view of the truth about the confrontation between Huxley and Wilberforce?

3 What historical evidence does he put forward to counter the view that all Christians were opposed to the idea of evolution?

4 Frederick Temple's viewpoint offers a way of believing in evolution and the existence of God. Explain how this is achieved. What do you think of this view?

5 What were the main influences on Darwin's religious faith?

6 How did Darwin's view on religion change throughout his life?

7 In 1925 a schoolteacher, Thomas Scopes, went on trial in Tennessee in the USA because he advocated the teaching of evolution in schools. What do you think were the main reasons behind the legal action taken against him?

2 LAUNCHING TECHNOLOGY

UNIT 2.1 The space race: competition or co-operation?

During World War II the USSR and the USA co-operated with other countries as part of the Allied Forces that eventually overcame Hitler's Germany. Yet this co-operation rapidly turned to hostility when the war ended in 1945. The two countries were each convinced that their own political system – communism in the USSR and capitalism in the USA – was superior. The next 40 years were an era of rivalry and suspicion described as the Cold War. It was a war in the sense that each side was desperate to prove its political superiority. This was manifested not by direct military confrontation but by competition in terms of economic, scientific and technological development. Two aspects of technology were particularly affected by this rivalry, and sometimes closely linked: the space race and the arms race.

When the Russian *Sputnik*, the world's first artificial satellite, was launched into an Earth orbit on 4 October 1957, it created a furore among the Americans who had previously thought that *they* were at the forefront of space technology. In April 1961 the USSR achieved another first when they launched a rocket into orbit with a human astronaut, Yuri Gagarin, on board. The Americans were again stunned by the news that appeared to show that they were trailing in the space race. There were some people who blamed the US education system for failing to produce enough scientists. To retrieve a sense of national pride the Americans set their sights on being the first to reach the next most important target in this race – a manned landing on the Moon. The gap between the first manned American space flight in June 1961 and the first landing on the Moon in July 1969 by the crew of *Apollo 11* was remarkably short. It had been achieved through considerable investment of human and financial resources. Some people saw it as a triumphal demonstration of the bravery and ingenuity of the human spirit. Others doubted whether the 25 billion dollars spent on the Apollo programme had been worth while.

The Russians admitted defeat in the race for the Moon after a number of setbacks, including the death of a cosmonaut, Vladimir Komarov, in April 1967 when his *Soyuz 1* space capsule hurtled to Earth out of control. They decided to advance their space technology in a different direction. In September 1970, an unmanned robot space craft landed on the Moon, scooped up a sample of Moon rock and returned it safely to Earth for scientific analysis. The Russians also gave priority to the development of manned space stations orbiting the Earth. Astronauts could live and work there for months at a time, conducting experiments and taking reconnaissance photographs of the Earth's surface. It was soon discovered that astronauts could endure lengthy space flights without experiencing life-threatening health risks. In December 1988 two cosmonauts returned to Earth after a record 365 days in space on board the latest Russian designed space station, *Mir* (see Figure 2.1.1).

With the thawing of the Cold War in the 1990s, it has been suggested that in future the US and the USSR should combine their efforts for any major space exploration project such as a manned landing on Mars. Co-operation between several European countries is already under way in the form of the European Space Agency (ESA), which regularly launches satellites using *Ariadne* rockets. A space plane that could travel from Britain to Tokyo in two hours is one of ESA's plans for the future.

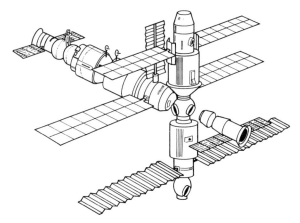

Figure 2.1.1   The Russian *Mir* space station.

ACTIVITY

Space missions – economic and political influences

1. Unlike manned lunar landings, interest in developing manned space stations has remained fairly constant over the last 25 years. Suggest why this might be.

2. Unmanned lunar landings are cheaper and pose no danger to human life. Why do you think the Americans chose to launch manned lunar missions instead?

3. Find out in what ways the early space flights of the Russians differed from those of the Americans.

4. a) Describe the technical, economic and political advantages of co-operation between countries in tackling major space missions such as a manned landing on Mars.
   b) What are the major obstacles to such joint ventures?

5. Imagine you are the President of the USSR and write a letter to the American President to persuade him to co-operate in such a joint venture.

6. As a class, debate the following motion with speakers for and against:
   - Competitiveness between countries is of vital importance in increasing the speed of scientific progress.

7. There have been no space flights to the Moon since 1974. What does this suggest about:
   - the cost
   - the military importance
   - the scientific importance of manned Moon landings?

8. American world prestige was boosted in 1969 when the USA achieved the first manned Moon landing.
   - Was this the prime motivation behind the *Apollo* programme? Or do you think that the desire to achieve greater knowledge was the main driving force?

9. The only rendezvous in space between a Russian and an American space craft occurred in 1975. It required years of joint planning to ensure compatibility of the two craft so that they could dock. What was the value of such a mission for scientific and political progress?

10. The passage below comes from an article written in 1966 by the Italian writer and chemist, Primo Levi, at the time of the *Apollo 8* mission, the first manned space flight to the Moon.
    - Discuss his arguments about the beneficial impact of the space programme.
    - Do you agree with him?

11. On the day of the *Apollo 11* launch, a group of protestors who were opposed to the waste and expense of the space programme marched to the gates of the launch site at Cape Canaveral. The head of NASA (National Aeronautics and Space Administration), Tom Paine, went down and spoke to them.

    > It will be a lot harder to solve the problems of hunger and poverty than it is to send a man to the Moon, but if it were possible not to push that button and solve the problems you are talking about, we would not push that button.

    - Discuss the assumptions behind this statement.
    - Do you think it satisfied the protestors?

The Moon and Man, by Primo Levi

Confronted by this latest evidence of bravery and ingenuity, we can feel not only admiration and detached solidarity: in some ways and with some justification each of us feels he is a participant. Just as every person, even the most innocent, even the victim himself feels some responsibility for Hiroshima, Dallas, and Vietnam, and is ashamed, so even the one least connected with the colossal labour of cosmic flight feels that a small particle of merit falls to the human species and so also to himself, and because of this believes he has greater value. For good or evil, we are a single people: the more we become conscious of this, the less difficult and long will be humanity's progress towards justice and peace.

From *The Mirror Maker*, Primo Levi

2 LAUNCHING TECHNOLOGY

UNIT 2.2 The *Challenger* Shuttle disaster

On 12 April 1981 a new era in space travel began. The world's first reusable space craft, the Space Shuttle, was launched in piggyback fashion attached to a massive fuel tank and two solid fuel booster rockets. Designed to fly like a glider as it re-entered the Earth's atmosphere, the Shuttle had retractable landing gear that allowed it to touch down on specially extended runways at about twice the speed of normal aircraft.

The Space Shuttle Programme had taken ten years and $4500 million of public investment to achieve its first launch. Over the next five years the public's imagination was captured by the sight of the Shuttle landing like an enormous, elegant bird. Its prestige was further enhanced when, on one flight in 1984, the astronauts carried out a delicate rescue mission in which they captured and repaired a damaged satellite using the Shuttle's robot arm.

But by 1986 the Shuttle programme had reached a crucial stage and its future was in some doubt. Politicians who opposed the programme were becoming increasingly concerned that the nine billion dollar space craft was placing an intolerable strain on the US budget. In the original plans it had been envisaged that the Shuttle would make 30 flights a year and become a profitable 'space taxi service'. Finance would come from carrying fee-paying passengers and commercial payloads such as satellites and scientific experiments. Because of numerous critical technical problems, only five successful Shuttle missions had been launched in the previous year. NASA now felt considerable political pressure to justify the enormous financial commitment that the American taxpayer was making. A dramatic increase in the number of launches would go some way to placating this dissatisfaction among the American public. In 1986 they set themselves the target of 15 launches within the following 12 months.

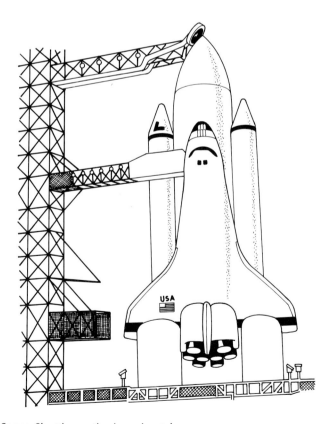

Figure 2.2.1   The American Space Shuttle on the launch pad.

The twenty-fifth Shuttle flight in January 1986 was therefore a particularly important mission for NASA. Of the seven astronauts in the crew of the *Challenger* Shuttle, most media attention was focused on the least experienced. She was Christa McAuliffe, an extrovert 37-year-old teacher and mother of two children, chosen from 11 000 applicants for the NASA 'Teacher in Space' project. She became well-known to the public through numerous interviews and announced that she would keep a journal of the flight, which she hoped would convey 'the ordinary person's perspective'.

The take-off at 11.38 a.m. on 28 January appeared at first to be going well. But only 73 seconds later, at a height of 13 000 metres, the Shuttle was destroyed by a massive explosion in the fuel tank, killing all seven astronauts on board.

It was the worst accident in nearly 25 years of manned space flight. The official investigation declared that the Shuttle flight directors had disregarded safety warnings from the engineers in their anxiety to meet the target of a certain number of launches per year.

FURTHER READING

The Nobel laureate, Richard Feynman, was a member of the committee that investigated the *Challenger* disaster. He gives a fascinating insight into the political and technical issues involved in his book, *What Do You Care What Other People Think?*, Unwin Paperbacks, 1990.

ACTIVITY

Shuttle problems

1 What was the general attitude of the American public towards the Shuttle programme at the end of 1985?

2 NASA had taken the unusual step of including the first private citizen – a woman who was both a mother and a teacher – on the Shuttle flight. What do you think NASA hoped would be the impact of this innovation on public opinion?

3 Before the launch of the ill-fated *Challenger*, engineers warned that the sub zero temperatures would be likely to affect the rubber O-rings that sealed the joints between the sections on the solid fuel rocket boosters and cause a critical accident.

   a) Suggest how the rubber might be affected by these low temperatures and lead to a critical leak of fuel.

   b) Devise an experiment to test the effect of sub zero temperatures on rubber.

   c) Suggest why the warnings were ignored by the NASA flight directors who decided to approve the launch.

4 Suppose you were applying to NASA to be a participant in the first 'Teenager in Space' project. Write your letter of application giving your reasons for wanting to be included.

ACTIVITY

To launch or not to launch – a role play

In groups of two, three or four try out a role play between the people involved in the conflict that led to the Shuttle disaster. Imagine it is the evening before the planned launch. The engineers are worried because the air temperature is lower than for any previous Shuttle launch. They decide to express their anxieties to the flight directors.

The rocket engineers
Because of the low temperatures you believe that a launch would be too risky and should be cancelled. However you can only give advice; the final decision is not yours.

The flight directors
You have had to cancel the launch twice in the previous week and you feel under pressure not to cancel again. The future of the entire programme may be at risk if the target number of launches is not maintained. You are determined to go ahead.

2 LAUNCHING TECHNOLOGY

UNIT 2.3 Future space exploration: a mission to Mars?

Sending astronauts to Mars could be a feasible project. But are we prepared to spend the vast sums of money that are required? In 1990 the American President, George Bush, declared that he wanted to see an American flag on the surface of Mars by 2020. If you were a politician elected to the Senate of the United States, would you vote to approve the expenditure needed to reach this goal? Or are there more pressing problems to be tackled, such as poverty, hunger and disease, that have a greater call on the world's limited resources of money and expertise? In this unit you will explore these questions by enacting an imaginary meeting of the Budget Committee of the US Senate. At the end of the meeting the Committee members must decide whether or not to fund a mission to Mars.

ACTIVITY

Should we send a mission to Mars?

The United States Senate consists of elected representatives, Senators, two from each one of the 52 States. A Committee of the Senate can call on experts to help the Senators to reach a decision.

Members of the class should volunteer for the seven roles described. Other class members can take on the roles of additional Senators on the Budget Committee. To make these roles realistic, decide for yourself your character's age, background, personality characteristics and attitude to the Mars mission.

Follow through the role play in four stages:
1 The experts give their opinion of the hazards of such a mission.
2 The Senators question the experts on their positions.
3 The Senators discuss what they have heard and give their own views on the mission.
4 A vote is taken on whether to go ahead with the mission.

Role descriptions

Senator Alex Hadlington
You have been appointed chair of the Committee. Your job is to call witnesses to give evidence, invite the other Senators to give their views and eventually supervise the voting. Your main task is to control the discussion so that everyone can make their points.

Senator Yvonne Galvin
You are a fierce opponent of the planned mission to Mars. You believe that the $100 billion budget would be a waste of money and should be spent on projects to relieve hunger and disease in the developing world and to boost medical research into treatments for incurable diseases such as AIDS, Parkinson's disease and cancer.

Senator John Lucas
You are a staunch supporter of NASA's aim of putting a human astronaut on Mars because you believe that this would lead to the formation of a base on Mars with the potential for human colonisation. Like the *Apollo* programme before it, the Mars mission would provide a test of human ingenuity and could generate valuable technological developments applicable on Earth.

Dr Rachel Zublick
Expert on solar radiation.

You have been called by the Committee to explain the hazards of solar radiation to which astronauts would be exposed during their journey through deep space between Earth and Mars. There is a high risk of a solar flare during a mission to Mars, which would emit a dangerous dose of protons. Like radioactive radiation, they could alter the DNA in human cells and increase the risk of cancer by about one per cent. The space craft would also be continuously bombarded with cosmic rays, which include high energy heavy particles such as iron ions, whose biological effects are poorly understood. Overall, the present state of knowledge does not provide definite evidence of the health risks of the level and type of radiation that would be experienced on a Mars mission.

Dr Paul Walstar
World-famous researcher at Harvard University on the effect of weightlessness on blood cells.

You have designed a number of experiments to discover the effects of long periods of weightlessness on animal cells. Animal experiments show a decline in the production of red blood cells that seems to worsen the longer the animals are in space orbit. You have also discovered that white blood cells taken from mammals divide very slowly when cultured in weightless conditions. This 'space anaemia' could lead to a serious deficiency in immune response to infections. However, the long-term effects are uncertain and the shortest mission to Mars would be 18 months, including 30 days exploring the planet surface. At a recent conference you summarised your views with this statement: 'Nobody has the foggiest idea of the effect of even 40 per cent or 20 per cent of gravity for extended periods. I expect people will want to go anyway, even though we may kill a few on the way, and we may get some wet noodles back.'

Professor Alice Weinberg
NASA's Chief Medical Scientist, employed since the first American space flights in the 1960s.

Your evidence to the Committee comes from your research on the effect of weightlessness on bone and muscle tissue of astronauts on their return to Earth from lengthy space missions. On the early flights, it was noticed that the masses of these two tissues were greatly reduced as a result of the lack of effort required to move in zero gravity. To counteract this effect crew members have used exercise bikes and treadmills with some success. However, this machinery is bulky and does little to provide the shock to bones of Earthbound exercise that they apparently need for health.

André Dupont
The first French astronaut to be launched into space.

You have spent 12 months on board the orbiting Soviet space station *Mir*, and have been called by the Committee to give evidence on the psychological problems of prolonged space flights. The tensions between the crew members in the cramped living quarters of the space station were so intense that they led to serious open hostility. The crew members were on the verge of mutiny and the mission had to be ended early. You also observed extreme restlessness and bouts of depression among fellow astronauts during this long mission on board the space station.

Science, Technology and Society © D. Andrews, 1992

2 LAUNCHING TECHNOLOGY

UNIT 2.4 Industry and innovation

To a much greater extent than in any other animal, the human species has always relied on its ingenuity to produce creative solutions to the problems of survival. This ability results from the tremendous powers of an intelligent brain, fine manual dexterity and the ability to share ideas through a spoken language. The earliest examples of human innovation, such as the use of fire and the construction of a flint-headed spear, expanded the opportunities for survival in hostile environments. Nowadays a constant flow of innovations contributes to improved living standards in many different ways.

For innovation to be successful, a variety of conditions must be met. An individual, or more often a company, is needed to bring together the following elements:

- Financial institutions and government funding agencies to provide long-term investment that does not insist on short-term returns. This type of investment is known as **venture capital**. With many inventions it may take decades for an idea to be converted into a commercially successful product.

- People with creative minds, such as scientists and engineers, who can apply new scientific knowledge to the development of ideas for products. This is called **Research and Development (R & D)**.

- Financial expertise to assess whether the innovation can be produced at an economic price. If it is too costly a good idea may not catch on.

- Marketing experts who will advertise and sell the product.

Having brought together these elements the innovation may still not succeed. There is still the vital question: will people want to buy the new product? Failing to predict consumer demand accurately has led to the commercial downfall of many innovations, such as the electrically powered three-wheeled vehicle, the Sinclair C5, launched by Sir Clive Sinclair, one of the pioneers of home computers, in 1985 (see Figure 2.4.1).

Another obstacle facing innovatory ideas is the legal framework that regulates industry. As an illustration of this problem, consider the national communication network that many scientists would like to see linking all homes and businesses. This innovatory system would be based on fibre optic transmission, which involves sending signals as pulses of a laser beam through very thin glass fibres. An integrated network could carry television, telephone, audio and computer data signals into homes and offices. However British government regulations would need to be changed for this innovation in telecommunications to be introduced.

Figure 2.4.1 Why was this technological innovation a commercial failure?

ACTIVITY

A national fibre optics network

Read the article below, which explores the potential advantages of a national fibre optic communication network. It also assesses the problems involved in introducing this innovatory system.

1 What is meant by:
   a) redolent   b) omnipotent   c) compatibility?

2 In what way is fibre optic cable an improvement on the conventional copper cable?

3 Explain what is meant by:
   teleshopping
   teleworking
   teleconferencing
   electronic freight transport.

4 What are the advantages and disadvantages of each of the above compared to the standard procedures that do not rely on electronic transmission?

5 What changes to the system of regulation would be necessary before the vision of a nationwide fibre optic communication network could be realised?

6 What have been the applications of the following innovations:
   holograms
   thermostats
   switches activated by breaking a light beam
   lasers?

Eyeing opportunities for a network

FIBRE OPTICS
● Political obstacles hinder communications

A dream of one omnipotent communications network, redolent with teleshopping, teleworking, a wonder called "electronic freight transport", as well as television, fax and telephone services, lies in the breasts of some British technologists.

But unless the government frees and rearranges its regulatory policy, this system cannot exist. Professor David "Den" Davies, vice chancellor of Loughborough University, told Science 89 that a fibre optic network could carry the mighty load far more quickly and cheaply than conventional cables. However the dominant role of British Telecom, coupled with the government's desire to promote competition in the communications field, is a huge obstacle.

Investment required for such a network would be around £20 billion. "We could obviously do it (technologically), but the problem area is the politics," says Davies.

One advantage of fibre optics is that great increases in capacity can be made at trivial costs. Improved immunity from interference and compatibility with existing networks are other pluses.

Davies compares the performance of a cable with a similar length of optical fibre. The cable could handle 97,000 speech channels, with analogue technology, at a diameter of 78 millimetres, with repeaters spaced 1.5 kilometres apart. Optical fibre would handle 300,000 channels digitally, with a diameter of 21 mm and repeater spacing of 30 kms.

Television was one perfect candidate for fibre optics, but "TV alone done by optics would not justify optical fibres," Davies says. There again thorny problems of regulation lie in the way. "It might work if BT and Mercury could carry TV but were not allowed to provide TV."

Telecommuting (working from home via a computer link), teleshopping and teleconferencing were all concepts, under experiment in many parts of the world, with obvious advantages. "Television business meetings may be just as effective as people travelling to meet up. And working at home may suit many people very well, although children pouring coffee down the backs of the machines may turn out to be a problem," he says. "As for teleshopping perhaps people will always prefer to walk around the shops."

Electronic freight transfer is the latest development, a move towards actual goods being transmitted electronically. The best example at the moment is newspapers, such as USA Today, which are in the most part edited and assembled at one site, then relayed electronically to satellite plants for printing.

"The problem awaits a satisfactory solution to the regulation of use of such a network," Davies says – falling short of making an actual plea to the government.

There are, he said, senior level discussions going on to find a wide role for fibre optics. "Many people are trying to find ways in which such a network could be introduced, but which didn't interfere with existing systems."

British Rail, which already has an extensive fibre optic network in place, has let it be known that it would be interested in capitalising on this service.

The *Sunday Times*, 17 September 1989

2 LAUNCHING TECHNOLOGY

UNIT 2.5 **Industry and the environment: bauxite mining**

At its simplest, industry is any enterprise that requires the close co-operation of a number of people. It is an activity that appeared very early in the evolution of the human species. Our distant ancestors needed to work with precise coordination to hunt fast-moving prey on the African plains. Nowadays, industry organised on a massive scale coordinates the efforts of thousands of individuals to tackle far more complex problems. Publicly or privately owned industries (see Box below) provide us with all the products and services that we need through the primary, secondary and tertiary sectors (see Table 2.5.1), which are closely dependent on each other.

While industry is vital for providing us with employment, it can also have a harmful effect on the environment. When a large industrial development is proposed, local people may react in very different ways. Some may welcome the new job opportunities it brings, while others may strongly oppose it. Opposition will be most vociferous if the new development will seriously reduce the quality of people's lives and damage the local environment. In this unit we will explore this issue through a case study of a fictional bauxite mining company, Prospectico International, that wishes to expand its operation in Jamaica.

Private and public industry

A private company is financed through bank loans or through the sale of shares to individuals. The owners of the company are its shareholders, who expect a share of the profits in the form of an annual payment (dividend).

Publicly owned industry is financed by national government from the money collected through taxation. The industry is owned by the government and operated in the interests of all the citizens of a country. It may well run at a loss, which is made up by a grant from the government known as a subsidy. For example, public transport systems in many countries receive government subsidies to minimise the cost to the traveller.

Table 2.5.1   The three sectors of industry

Industrial sector	Function	Example
Primary (Extractive industry and agriculture)	Extracts the Earth's natural resources to provide raw material for the secondary sector	Coal mining
Secondary (Manufacturing industry)	Creates products from raw materials	Car manufacture Food processing
Tertiary (Service industry)	Provides services rather than products	Tourism Education Catering

ACTIVITY

Sectors of industry

1 Classify the following organisations according to whether they:
   a) are privately or publicly owned
   b) belong to the primary, secondary or tertiary sector.

   Nuclear Electric
   National Power
   The National Health Service
   British Coal
   Post Office
   Harrods department store
   Your nearest swimming pool
   Your school/college
   British Telecom
   British Rail

2 It has been suggested that the subsidy given to public transport in the UK should be greatly increased, reducing the fares so that more people will use it rather than take their cars. Other people have argued that transport systems should be handed over to the private sector. Which do you think is the better option?

DO WE WANT A BAUXITE MINE?

Minerals found in the Earth's crust provide valuable raw materials for manufacturing industry. The mineral bauxite, for example, is processed to form aluminium, a metal with many important properties, including good resistance to rusting, which makes it suitable for manufacturing a huge range of products, from kitchen foil to space rockets.

Bauxite is mined by the open-cast method, which is used whenever minerals are found close to the Earth's surface. This method is relatively cheap compared to deep seam mining, but it has a particularly devastating impact on the landscape. Instead of boring shafts and extracting the mineral from seams deep underground, the top layer of soil is removed to expose the seam on the surface.

ACTIVITY

A bauxite mine for Saint Francis?

Imagine you are a resident of the imaginary town of Saint Francis on the Caribbean island of Jamaica. A bauxite mine and treatment plant have been proposed on the outskirts of the town and the residents are debating fiercely whether this is a desirable development. The opposing viewpoints are expressed in the two documents that are reproduced here. The letter has been sent to every resident of Saint Francis by the chairman of Prospectico and the leaflet has been produced by the campaign opposing the mining development.

1. What do you think would be the different views on the proposed mine of the following people:
   a) a teenager
   b) the owner of a small store
   c) an elderly person?

2. As a class, work on a role play to dramatise the conflicting viewpoints. You might want to stage a role play of a meeting where the Chairman of Prospectico and the leader of the opposition campaign share the same platform. They each put their case to the audience and then answer questions.

3. After the role play, write an account of the meeting as a front page article for the local newspaper.

Science, Technology and Society © D. Andrews, 1992

PROSPECTICO INTERNATIONAL

ALUMINA AVENUE • NEW YORK 543762 • USA

21 May 1991

Dear Resident

Saint Francis desperately needs new industry to breathe life into its sick economy. Prospectico International urges you to support the opening of a bauxite mine.

- *Unemployment is driving the young people away from the town and into the cities. The new mine will give them the chance of local employment and help to revitalise the town.*
- *Transport to and from the town has always been difficult because of the poor roads and unreliable bus service. The company will improve the road linking Saint Francis to the capital and contribute to the cost of a new railway line.*
- *Prospectico International has always had a responsible attitude to the environment. We will ensure that pollution from the mine and treatment plant is kept as low as reasonably possible while maintaining a commercially viable operation.*

I trust that in the next few months while we wait for government approval we can satisfy any concerns that you have.

Yours faithfully

B.J. Oreshoveller

B.J. Oreshoveller
President, Prospectico International

SAVE SAINT FRANCIS
STOP THE BAUXITE MINE!

The new bauxite mine would be a disaster for the residents of Saint Francis and for the local environment. We urge you to support our campaign to stop this ugly, destructive and pointless development. We argue:

- The planned development will destroy the peace and tranquillity that we value here.
- The mining process will involve massive noisy digging machinery and huge quantities of dust will be produced. The fumes and noise will be dangerous to our health.
- The treatment plant will pollute the water in the local river, killing the fish that live in it.
- Vast areas of rainforest will be destroyed and the species that depend on it will perish.
- Why should we suffer to produce a metal that the richer countries are throwing away in huge quantities? They should be recycling aluminium, not destroying our environment to get more of it.

JOIN OUR CAMPAIGN TODAY

3 INFORMATION TECHNOLOGY

UNIT 3.1 Trends in information technology

Information technology (IT) includes the collection, processing, storage and communication of data in all its forms – pictorial, verbal, written or numerical. One of the characteristics that is uniquely human is our sophisticated ability to store and transmit information using a spoken and written language. It was crucial in the early stages of human evolution because it allowed us to discuss our hopes and fears with other members of our species and to plan for future co-operative action. As technology has advanced, our capacity for communication has been extended in many ways and this has had a significant impact on the way society is organised. One recent advance in communications technology has been the switch from analogue to digital transmission of signals (see Box 2). One application of this innovation has been the switch from vinyl records to compact discs.

1 COMMUNICATION MODES

- fax
- face-to-face speech
- television
- telephone
- visual signalling – smoke signals – flags – flashing lamps
- telex
- radio
- electronic mail from one computer terminal to another
- wireless telegraphy using morse code
- letters transported by – hand – horseback – train – air

ACTIVITY

Communication technology

If you want to deliver a message to someone you can use one of the communication methods listed in Box 1. Carry out your own research on these different technologies, using the resources of a library to help you answer the following questions.

1 Arrange the communication methods given in order of the approximate date of their invention.

2 Describe the main advantages and disadvantages of each of these communication methods available today, taking account of the following factors: cost, speed of delivery, speed of reply, accuracy of the transmitted message, flexibility of possible modes that can be transmitted (verbal, written, pictorial, numerical), vulnerability to noise interference (Box 2).

3 How did the following innovations in communications technology influence society and relationships between people:
   a) the telephone
   b) television
   c) satellites?

4 Imagine that you wish to communicate a vivid description of your daily life to a person living in another country. Compare the advantages of using the different methods listed in Box 1 to communicate this information.

5 What does Figure 3.1.1 tell you about the computer industry between 1980 and 1986?

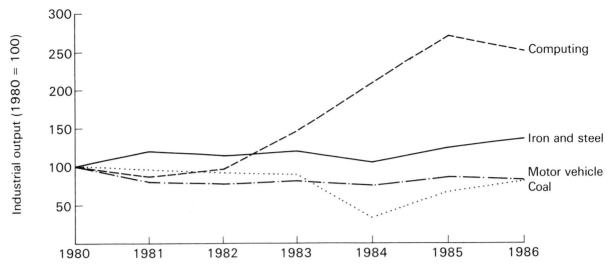

Figure 3.1.1 The annual output of different types of manufacturing industry in the UK between 1980 and 1986.

② Noise and the mode of transmission

To communicate your message clearly using any mode, you need to ensure that it is not distorted or hidden by irrelevant interference or 'noise'. The crackle on the telephone line or the white lines on a videotape are examples of noise. If it becomes excessive, noise will swamp the signal, making it impossible to understand the message. In communication systems it is vital that the ratio of noise to signal is carefully controlled. Some modes of transmission have a superior noise/signal ratio to others.

Electrical or optical signals can be transmitted in two different modes – digital and analogue. Analogue signals vary continuously over time and are transmitted in the form of waves. Digital signals code the same information using binary code, so the signal can have a value of either 1 (on) or 0 (off). (See Figure 3.1.2.) The main advantage of digital transmission is that it is much less vulnerable to the distorting effects of noise. With an analogue signal, any noise is not distinguished from the signal itself and so they are amplified together. Record players that carry analogue signals reproduce all the scratches on vinyl records, but CDs cannot be scratched and digital CD players do not pick up fingerprints so they are 'hiss-free'.

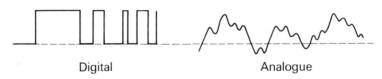

Digital   Analogue

Figure 3.1.2 Digital wave forms have discrete values of one or zero. Analogue wave forms vary continuously.

3 INFORMATION TECHNOLOGY

UNIT 3.2 From valves to transistors to microchips

Electronics is a young science. Yet in its short life it has already gone through three major revolutions, each of which relied on the invention of a new component:

- the triode valve in the 1920s
- the transistor in the 1950s
- the microchip in the 1970s.

These components all perform a similar function: the amplification and switching of signals in radios, televisions, computers and a host of other electronic devices. But the advance that each one made on its predecessor paved the way for major technological innovations that had widespread effects on society (Figure 3.2.1).

The valve allowed the development of radio and TV broadcasting in the 1920s and 30s. By the time World War II broke out everyone had access to radio broadcasts, which for the first time gave politicians the opportunity to talk to the entire population. Using radio receiver-transmitters, people living in isolated and hostile territory could be contacted. This helped in the provision of medical and educational services to remote regions such as the Australian outback.

The early valve-driven computers filled entire floors of buildings and generated a great deal of heat. Also, like a light bulb, the valve filament only had a limited life, so the computers were unreliable and could only run short programs. What was needed was a smaller, more reliable component that could do the same job. The invention of the transistor (Figure 3.2.2) in 1948 by the US scientists Bardeen, Brattain and Shockley started the miniaturisation trend, which is continuing today. In the USA and the USSR the pressure of the arms race (see Unit 4.5) and the space race (see Unit 2.1) added to the drive towards ever smaller, faster and more powerful computers that could be squeezed into the nose cones of missiles or space craft.

The next breakthrough in miniaturisation came with the invention of the integrated circuit in 1960. The equivalent of thousands of transistors could be etched on to wafer thin slices of silicon only 5 millimetres square. The computing power of the 1944 valve-driven Harvard Mark 1 Computer, which was 16 metres long and 1.5 metres high, could now be compressed into a single microchip. The silicon microchip could also be

Figure 3.2.1 Each stage of the electronics revolution resulted in miniaturisation.

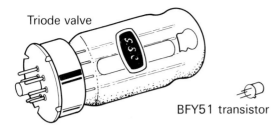

Figure 3.2.2 During the 1950s the smaller, more reliable and more power efficient transistor rapidly replaced the triode valve.

mass produced using a photographic process, and this opened up the possibility of the desk top computer for use in offices and homes.

Since its introduction, the microchip has been further miniaturised (see Figure 3.2.3) – a process that is still continuing today.

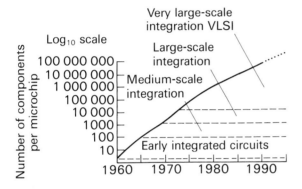

Figure 3.2.3 Graph showing the increase in the number of components per microchip (1960–1990).

Table 3.2.1 Landmarks in computing

Date	Development
pre 3000 BCE	Fingers used as the first calculating aid for counting in multiples of ten.
1300 BCE	First appearance of the Chinese abacus.
AD 1642	Pascal develops the first mechanical calculator. Clerks and accountants fear for their jobs, but few calculators are sold.
1859	Charles Babbage designs an Analytical Engine based on a complex system of cogs and levers. It has all the features of a computer – a memory store, a calculating unit, and input/output devices. When built it fails to work.
1890	Herman Hollerith wins a competition to design a system for counting the US census. Information is coded on punched cards and sorted at high speed using electrical power.
1944	The first generation of computers is launched with the Harvard Mark 1. It is a valve-driven digital computer, based on electromechanical relays.
1956	The second generation of computers, based on transistors, with vastly improved reliability.
1960s	The third generation of computers, based on silicon chip integrated circuits. The desk top computer becomes a possibility for the first time.
1980s	The fourth generation. Very large scale integration (including more circuits on each chip) allows improved miniaturisation and processing speeds.
1990s	The fifth generation. Computers with artificial intelligence achieve new levels of problem-solving ability, e.g. the translation of speech input into text output and vice versa.

ACTIVITY

Advances in communications

1 Give three ways in which the transistor is an advance on the triode valve.

2 The era of mass communication started with the invention of radio. How did this technology change the way politicians conveyed their policies? Find out what the wartime leader Winston Churchill owed to radio.

3 Discuss these statements:

'Without the arms race and the space race, the rate of progress in microchip technology would have been much slower.'

'One of the main harmful effects of television has been its contribution to the decline in live entertainment and socialising outside the home.'

4 What is likely to be the future trend in the graph shown in Figure 3.2.3?

3 INFORMATION TECHNOLOGY

UNIT 3.3 Computer applications

The inventors of the early computers saw them simply as complex calculating machines. But in the early 1950s people began to realise their possible applications. This unit explores how large and small businesses have been able to exploit computers.

ACTIVITY

Using computers

Consider the following activities.

- Newspaper and book publishing
- Supermarkets
- Banks and building societies
- Education

For each one, answer the following questions.

1. How have computers been used?
2. What has been replaced by the use of computers?
3. What benefits have been gained by:
   a) the consumer   b) the organisation?
4. What problems have arisen because of the introduction of computers?

ACTIVITY

CAD-CAM

Computer Aided Design (CAD) allows designers to build up a drawing or plan by adding lines and components through the keyboard. Computer Aided Manufacture (CAM) is applied in a wide range of industries from car manufacture to biscuit making. The application of a combination of CAD and CAM to the design and cutting of yacht sails is described in the article below. Use the information in this article to answer the following questions.

1. Explain what is meant by
   a) a skill shortage
   b) production bottle-necks
   c) traditionally based industries.

2. Suggest why traditionally based industries are more likely to be intimidated by CAD-CAM compared with high tech businesses.

3. What were the three main attractions of using CAD-CAM for Kemp Sails' production process?

Plain sailing with CAD-CAM

It is becoming more important for small businesses to be aware of the benefits of techniques such as computer aided design and manufacture in overcoming skill shortages and other production bottle-necks.

The advantages are already recognised by technologically based businesses, but it is those firms that are more traditionally based that can probably benefit the most. They are also probably the most intimidated by the prospect of wrestling with computers. An indication that the situation is changing can be gained from the growing number of entries in the annual *Livewire* programme which aims to encourage young people between the ages of 16 and 25 to create their own self-employment. Many of them show a clear understanding of the potential benefits of CAD-CAM.

Inevitably most entries for the award still lean towards word-processing and database management uses, which, while they have value, are not necessarily the most important application for the young business. However, in a large number of entries this year an understanding of the wider benefits came through clearly. Rob Kemp, of Kemp Sails in Dorset, won the *Guardian*'s £1000 computer award because he showed an awareness of how CAD-CAM could enable him to expand his business in the traditional trade of sail-making.

Mr Kemp, aged 23, started his business just over three years ago. He now provides full-time work for four people and part-time work for several others, and is developing exports to Europe and the USA.

He faced the fact that as the business grew it would be difficult to find and keep skilled staff. 'The main advantage of CAD is that we can save a lot more floor space in our loft. By installing a CAD-CAM system we can employ untrained staff and then train them to do the job our way. This in turn gives us increased stability and security. It also gives us a tremendous marketing advantage since the America's Cup and TV series like *Howards Way* have shown yachting and yacht racing as high tech industries. Our customers expect us to have the computer facilities to offer them computer cut sails.'

UNIT 3.4 Artificial Intelligence

Soon after their invention, computers showed their value as powerful tools for storing, sorting and manipulating data. Then many people began to realise that their potential was much greater than this. They saw that the computer could apply a set of logical rules to solve more complex but rigidly defined problems, such as playing chess. When a computer beat a human chess champion the newspaper headlines proclaimed that an electronic brain superior to a human's had been invented. Once the chess expert had become familiar with his or her opponent's strategy a draw could easily be forced, but very rarely a win.

In the 1980s researchers began to develop computer software that could tackle problems using more complex arrangements of rules in a more flexible way, much as we use our own intelligence. A number of different Artificial Intelligence (AI) techniques, as they are known, are rapidly being developed and many people are asking the question, 'Can computers develop an intelligence that matches or even improves on the human brain?'

EXPERT SYSTEMS

These systems are developed by interrogating human 'experts' and analysing the steps they take in tackling a problem. These steps are then translated into a system of rules that are written into a computer program, which will then solve the problem in the same way. For example, a doctor reaches a diagnosis by asking a series of questions about the patient's symptoms.

Expert systems designed to tackle a simple problem using a few hundred rules are often very effective. But larger systems come across a number of problems. People find it difficult to describe exactly the rules of thinking that help them to solve very complex problems. Hunches, guesses and flashes of inspiration are all valuable, yet they are not capable of systematic description. Expert systems get round this difficulty by ascribing a certain mathematical probability to a rule. This approach has the drawback that the solution to the problem cannot be given with complete certainty.

A recent advance in expert systems allows them to work out the rules themselves by processing a number of examples. This technique has been used to construct a program that decides whether an applicant for a credit card should be given one. Researchers fed in information about a large number of applications and the decisions taken when they had been processed by humans. By analysing these data the computer was able to 'learn' the appropriate rules to apply.

NEURAL NETWORKS

One obvious approach to writing software that mimics human intelligence is to try to copy the structure and processes that actually occur in the network of nerve cells (neurones) in the brain.

This technique is being used mainly in visual recognition. Scientists studying animal brains have built up a basic understanding of how the nerve cells are connected together to allow us to recognise shapes. Computer systems based on similarly designed neural networks are now capable of recognising different banknotes, for example, or faces in a company's security system.

GENETIC ALGORITHMS

Natural selection operates on living organisms by generating a large number of possible 'solutions' to the problem of survival. Of the many random mutations that are produced, only a few are successful in improving the animal's survival chances. These 'solutions' are selected and survive to produce the next generation (see Unit 5.1).

Genetic algorithms are a new AI technique that tackles problems in a similar fashion. First a large number of possible solutions are generated. The best ones are selected and 'mated' to produce several offspring solutions. The best of these are again selected and allowed to pass into the next generation. Repeating this process over several generations eventually improves the solution. This technique is currently being used by NASA to help design an orbiting space station.

ACTIVITY

The potential of AI

1 How far do you agree with the following statements:

   'A computer is just an advanced calculating machine combined with a vast library of information.'

   'We will eventually develop computers that have an intelligence superior to our own.'

2 Suggest why an expert system that can help to diagnose illness would be valuable in the following situations:

   on a manned space flight to Mars
   in a remote region of a developing country.

3 The plot of the short story *The Human Operators* by A. E. Van Vogt is summarised opposite. It gives a vision of the eventual domination of human action by intelligent computers. Do you think that there is any real danger of this happening in the future?

4 In Glasgow an expert system has been devised to help in the diagnosis of abdominal pain. How would you feel if the task of asking questions about your symptoms was delegated to a computer?

The Human Operators, by A. E. Van Vogt

To explore the remote regions of the solar system, humans have built specially designed space ships. On board each one is a computer with high powered AI capability and a single human 'operator' to carry out routine maintenance.

Once the fleet of space ships has been launched the intelligent computers override their original instructions and set off to explore deep space beyond the outer reaches of the solar system.

Very soon the limited life span of the human operators becomes a serious problem for the intelligent computers. When the operators die the computers could fail because their malfunctions would not be repaired. The computers solve this problem by ensuring that the space ships meet up at intervals to allow the humans to mate and produce a new generation of maintenance workers. Over the millenia the human operators carry out their tasks unquestioningly in the service of the intelligent space ships as they continue their wanderings through space.

3 INFORMATION TECHNOLOGY

UNIT 3.5 The Data Protection Act: who is holding information about you?

Many organisations store vast amounts of information about you and millions of other people. These stored personal data build up over the course of your life as you come into contact with more organisations. Every time you fill in a form, to open a bank account for example, the personal data you disclose are available for the organisation to use in any way it regards as appropriate, which can include passing on the information to another organisation. Information has traditionally been kept in written form, but nowadays it is more likely to be stored and transmitted electronically in the form of a computer file. This new means of storage and communication allows information to be accessed quickly and transferred easily from one computer to another. When confidential personal data is involved there is always a need to protect people's privacy. This need raises several important social issues that are addressed in this unit.

COMPUTER SECURITY AND COMPUTER HACKERS

The retrieval of personal data from computer files should only be carried out by someone who is authorised and has a justifiable reason. You want to be sure that your personal data are protected from the prying eyes of someone who is merely curious. Even more important, you want to prevent your personal data from being disclosed to an unscrupulous person who might use them against you. With written files this security is provided by locking them in filing cabinets with keys, to which only certain people have access. In a similar way computer files are 'locked' by security coded passwords. Access to a particular file is only possible if the correct security code is used.

A computer security system designed to protect personal data can never be 100 per cent foolproof. With sufficient time, skilled computer experts – hackers – can break into private computer files. Hackers relish the exhilaration of solving the complex maze of security codes that are intended to block unauthorised access.

The ingenuity of certain highly skilled hackers has enabled them to gain access to secret information about the US space programme from the NASA computer system.

ACTIVITY

Computer security

1 What organisations do you think store personal data about you at present?

2 How did these organisations obtain these data?

3 What might indicate that these data are stored on computer?

4 Suggest the possible advantages of storing records in computer files compared to written files.

5 In your view, which system of storage, computer or written, is most secure from unauthorised access?

6 Educational institutions store certain personal data about their pupils for a variety of reasons.
- In what form are these personal data stored in your school or college?
- Find out what type of information about pupils is stored.
- How is access to these records made secure?
- Who has access to these records?

7 Some hackers reveal their methods to the organisations whose computers they have gained access to. They claim that this is a useful service, which gives the organisation the opportunity to improve their computer security. Do you agree?

8 Should computer hacking be made illegal?

YOUR RIGHTS OVER COMPUTER STORED DATA

Computer storage of personal data is prone to serious errors. The information could be entered wrongly or be out of date. Personal data for one individual may become confused with those of another. This could lead to the person being refused credit, or a job, or even being wrongfully arrested.

Politicians reacted to this potential danger by introducing new legislation. The Data Protection Act, which came into effect in 1987, gives you the following rights:

- the right to know if any organisation holds computer records about you
- the right to see a copy of these records
- the right to have this information corrected if it is inaccurate
- the right to compensation if you suffer because of inaccurate personal data held about you
- the right to complain to an official body, the Data Protection Registrar, if you do not like the way organisations are using personal data about you.

However, the Act only applies to computer records; it does not cover written records. Certain types of personal data are exempt from the Act. For example, you would not be given a copy of your personal records if the organisation believed that this would hinder the detection of a crime. In most circumstances you have the right to see your medical and social work records, but you could be denied access to them if your doctor felt that seeing them would harm your state of health.

ACTIVITY

The Data Protection Register

Your local public library will have a copy of the Data Protection Register. This lists all the organisations that hold personal data.

1 Ask at the library for a copy of the index to the Register and pick out three organisations that may hold personal data about you. Consult the entry for each of these organisations in the main Register (held on microfiche). Use the information contained in the entries to answer the following questions about each organisation:

- What items of personal data does it hold on you?
- On what groups of people does it hold data?
- For what purposes does it hold these data?
- From what sources does it obtain the data?
- To which organisations does it disclose data?

2 Share this information with the other members of your class. Which organisations are likely to hold information on all class members?

3 Write a letter to one of your three chosen organisations asking for a copy of the data it holds about you.

4 Could your name and address have been transferred to the mailing list of another organisation? If so, how could you get it removed from the list?

Science, Technology and Society © D. Andrews, 1992

SCIENCE TECHNOLOGY AND SOCIETY 3 **INFORMATION TECHNOLOGY**

UNIT 3.6 Could a computer do your job?

When computers are introduced into a factory or office the employees' reactions are often mixed. There may be anxiety about the possibility of redundancy or having to learn new skills. But there may also be excitement in the novelty of new technology and the hope that it will make the jobs more interesting or easier to do. In this unit you will explore the feelings aroused when the introduction of new technology puts people's jobs at risk.

ACTIVITY

Cutting jobs in the salaries department

Box 1 gives the conversation between an office manager and a systems analyst. Two volunteers could act out this scene for the rest of the class. Then discuss the following questions.

1. The growth of IT has created many new jobs in the computer industry and has expanded opportunities for working from home using 'teleworking' technology (see Unit 2.4). But at the same time it has destroyed millions of jobs in more traditional areas such as clerical work and manufacturing.
   - List jobs previously performed by humans that can now be carried out by computers and robots.
   - List jobs that have been created by the growth of IT.

2. In what jobs have computers made the job holders more efficient without replacing them completely?

3. Suggest some jobs where there is future potential for:
   - the replacement of humans by computers
   - the improvement of human performance with the aid of computers.

4. Are there some jobs that you think could never be done by computers or robots? Could they do the job of a teacher or a surgeon, for example? Give your reasons.

5. What two features of computer technology does the systems analyst mention to emphasise its superiority to human workers?

6. How well do you feel Ms Allard handled this tricky situation? What do you think she has learnt from this confrontation with Mr Ramsey?

7. Discuss the points made by Mr Ramsey.
   - Are they good arguments to prevent the introduction of computer technology into his department?
   - Can you think of any better arguments he could have used?

8. Suggest why Mr Ramsey reacted in this way. Do you think that many people share his feelings about computers?

9. It has been predicted that the growth in IT will mean that 20 per cent of people will be working from home by 2010 and the average working week will drop to 34 hours. How will these trends affect our attitudes to:
   - the home
   - work
   - leisure?

10. 'Computers and robots are not doing just the boring and dangerous jobs for us. They are destroying all our jobs and the right to work.' Discuss this statement.

1 The visit of the systems analyst

Katherine Allard and Eric Ramsey both work for Everglow, a small company that manufactures light bulbs.

Katherine Allard is 23 years old and joined the company as a systems analyst two years ago after graduating in Business Information Technology. Her job is to suggest how the efficiency of departments in the company could be improved by the introduction of computer technology. She enjoys her work despite occasionally feeling anxious about the responsibility of her position in the company.

Eric Ramsey is 48 years old. He left school at the age of 16 and immediately joined Everglow as a junior clerk. He has worked his way up to the responsible position of salaries manager.

Time: 11.00 a.m.
Place: Mr Ramsey's office.

Katherine Allard: Good morning, Mr Ramsey.
Eric Ramsey: Good morning. [They shake hands and sit down.]

KA: As I'm sure you know Mr Ramsey, at the present time all our competitors are going flat out to cut their costs by bringing in computer technology. If we fall behind, the company could get into serious financial difficulties. So I'm here to have a look at the system you are using for calculating the staff salaries.

ER: [Anxiously] The boss hasn't been complaining has he? We hardly ever make a mistake in this department. I make sure of that.

KA: No, no, not at all. The managing director has said you are a credit to the company. All he wants to do is to see whether we can improve the efficiency of the salaries department.

ER: How can we do that? We're as accurate as humanly possible. OK, we make the odd error, but you've got to expect that occasionally. We're not robots!

KA: Yes, with 500 employees we would expect to have one or two of them finding mistakes in their salaries each month. We can't hope to eliminate human error completely – no matter what system we use. However we can speed up the time you take to do all the calculations. Working out each employee's salary deductions *is* very time-consuming, you must agree.

ER: But we've made a lot of progress since I started work here 32 years ago. We didn't have electronic calculators then like we have now. In those days there were 15 people doing the work of the five staff I've got here now.

KA: Well I think we could take this progress even further. Just one microcomputer could do the necessary calculations and print out the salary slips for all our employees. And it could give the Board members accurate and up-to-date figures on the salary costs of the company.

ER: Yes... but I do that already.

KA: And how long does it take you?

ER: [Proudly] I can get my monthly report compiled inside three hours.

KA: Mr Ramsey, you take three hours to write a monthly report. The computer could do a 100 per cent accurate report in a few minutes.

ER: [In an increasingly desperate tone] Well... OK, it may be quicker... and it might not make mistakes... but it can't do everything we can... what about an employee who rings up to complain about her salary deductions? I'd like to see a computer that could handle that!

KA: Yes, I agree. We will still need one member of staff here to resolve that sort of problem.

ER: One member of staff! One person is to cope with the salary administration for all the company's employees? It's impossible!

KA: [Getting up from the chair] Well, I'm afraid that whatever your view, the final decision is not yours, Mr Ramsey. I'll make my recommendations to the Managing Director and he will decide.

[Katherine leaves the office.]

Key roles in the company

Managing director
You have been appointed by the shareholders to coordinate the work of the employees and ensure that the company is successful. Your success is judged by the company's profitability.

Systems analyst
Your job is to analyse the tasks performed by the various company employees. You are required to cut the costs of the company by replacing tasks normally done by humans with the latest computer technology.

Trade union representative
You have been elected by your members to argue their case in any disagreement with the management of the company. You are keen to avoid any of the employees being made redundant.

ACTIVITY

Computers and jobs

The following series of improvised scenes will help you to appreciate the feelings of people faced with the introduction of computer technology. Class members should take on and act out the roles of different employees within an imaginary company. Three of the key roles are described in Box 2.

1 Decide on a name for the company. What does it produce?

2 Allocate the three roles in Box 2. The remaining members of the class should choose a specific job within the company for their role, for example: draughtsman/woman, accountant, engineer, clerk, advertising manager, secretary, shop floor production worker.

3 To help make your role more realistic, decide on the following:
- What is your character's name?
- How old are you?
- How long have you worked for the company?
- What is your present family situation?
- What does your job involve?
- How much do you enjoy your job?
- What is your attitude to computers?

4 Act out the following sequence of improvised scenes. After each scene discuss the questions that follow.

Scene 1 Setting the scene – the company at work

Decide on an appropriate location for the opening scene such as an office, workshop or laboratory. Arrange yourselves in appropriate groups according to your chosen jobs and role play conversations where you are carrying out your job. It is 9.00 a.m. on a Monday morning.

What extra information about the company do you need to agree to help you develop a more realistic improvisation?

Scene 2 The systems analyst starts work

Continue the improvisation from where you left off in Scene 1. After a few minutes the systems analyst enters the room and goes around asking the employees questions to help decide whose job can be replaced by computer technology. For example:

- What does your job involve?
- How does your job contribute to the success of the company?
- What do you achieve in a week?

1 What did it feel like to be questioned by the systems analyst? How did you react?

2 How did the systems analyst cope with the difficult task of asking delicate questions?

Scene 3 The managing director and the systems analyst

In this scene the systems analyst reports back to the managing director on his/her initial discussions with the employees. They also discuss the next step in the rationalisation programme and tackle the following questions:

- In which jobs will computers significantly reduce the workload?
- What can the company offer employees who lose their jobs, e.g. retraining?
- How will their decisions be communicated to the employees?

Scene 4 The union meeting

The trade union representative holds a meeting with the employees to discuss the rationalisation programme. It is an opportunity to voice views and anxieties about the possibility of redundancies and to decide on any action to take to avoid them.

1 What feelings were expressed in this scene?

2 How have people's attitudes changed since Scene 1?

3 How did the union representative react during the meeting?

Scene 5 Negotiating an agreement

A meeting is held to see if the two sides can come to an agreement on the introduction of new technology. The managing director, the systems analyst, the union representative and three employees are present.

1 What did the negotiations reveal about the difficulties of introducing new technology?

2 Draw up a set of recommendations for companies trying to introduce new computer technology.

3 Draw up a set of guidelines for trade unions negotiating with companies who seek to introduce new technology.

3 INFORMATION TECHNOLOGY

UNIT 3.7 Marketing innovations: satellite television

In 1962 the first live television pictures to be relayed using a satellite were broadcast to millions of viewers. The satellite, Telstar, received signals from the USA and beamed them down to the 76-metre diameter dish at Jodrell Bank in Britain.

Twenty five years later, the advance in telecommunications technology was such that a dish only 1 metre across could receive signals transmitted directly from an orbiting satellite. This opened up a new era of television broadcasting in which people could enjoy the benefits of space age technology in their own homes. But as with many technological innovations, satellite television's early days were hampered by financial problems. Any company launching a new product must be aware that there are considerable financial risks involved. When the product demands a huge investment in expensive and untried technology the uncertainties are even greater. Such was the case with the companies that launched satellite television stations in the late 1980s. These companies needed favourable answers to a number of questions if they were to make a financial success of this radical innovation in broadcasting technology:

- Can we attract enough consumers to purchase the equipment for receiving satellite television transmissions?
- How much will the consumer be prepared to pay to acquire the product?
- Can we gain sufficient revenue through the sale of advertising air time to finance the business?
- Can we keep down the costs of setting up a television station and satellite transmission system?
- Can we produce good quality programmes on a low budget?
- Is there a danger that the competition from similar companies will be so strong that we cannot attract sufficient subscribers?

In 1988 the first UK satellite television station, Sky TV, was launched. The second satellite television company, British Satellite Broadcasting (BSB), began broadcasting in May 1990, but failure to attract subscribers and advertising revenue soon led to financial problems. In order to survive, BSB had to merge with Sky in 1991.

The BBC and ITV networks broadcast their signals through land-based television transmitters on VHF

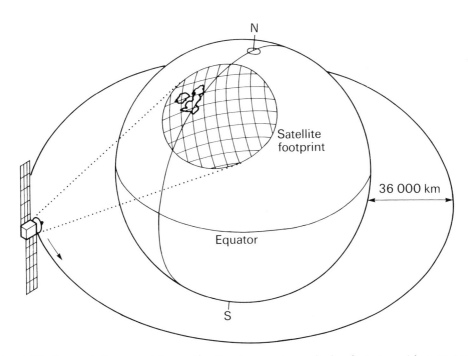

Figure 3.7.1 A satellite in geostationary orbit over the Equator can transmit signals over a wide area. Use this diagram to explain the orientation to the horizon of domestic television satellite dishes.

(very high frequency) and UHF (ultra high frequency). Unlike medium wave or long wave transmissions, these high frequency signals cannot pass through solid objects such as hills or large buildings. They can normally be picked up only when the receiving aerial is aimed directly at the transmitter mast. To cover the uneven terrain of even a relatively flat country the size of Britain, several hundreds or possibly thousands of transmitter masts, some of them over 300 metres high, are needed. The cost of construction, repair and maintenance of these transmitter masts is a massive slice of a television company's budget.

The advantage of a satellite is that its single aerial can transmit a television signal to a vast area – its footprint – which can be as large as an entire continent (Figure 3.7.1). The satellite must be parked in a special orbit 36 000 km above the equator, travelling at over 11 000 km per hour to match the speed of the Earth's rotation. This geostationary orbit ensures that the satellite remains in an unchanging position relative to a point on the Earth's surface. This principle was first devised in 1945 by a scientist who later became famous as a science-fiction writer, Arthur C. Clarke.

It is much cheaper to set up a satellite transmitter system than one based on land and this makes this type of system attractive to developing countries that might want to expand their national television network into remote regions.

ACTIVITY

Will satellite television be a success?

1 Devise a questionnaire among students at your school or college to investigate their experience of and attitudes to satellite television technology. Some of the questions you could explore might include:
- What proportion have ever watched a satellite television transmission?
- What proportion have direct access to satellite television transmissions in their own homes?
- What are their views on the quality of satellite television programmes?
- Do they regard satellite television dishes on the walls of houses as eyesores?

2 Discuss whether satellite television will become as popular as land-based television in the future.

4 WAR, TECHNOLOGY AND SOCIETY

UNIT 4.1 Warfare and technological development

It is a sad fact that throughout history, from the first flint axes to the atomic bomb, human societies have sought to enforce their authority over each other with increasingly advanced weapons. In this unit you will investigate further examples of the influence of society's war requirements on the speed and direction of scientific research and technological innovation.

Perhaps the most crucial technological advance was the replacement of the sword by the gun. For the first time combat between two armies could occur without soldiers coming into direct face-to-face contact. Over the centuries, with the development of more sophisticated guided missiles, the gap between the soldier firing the weapon and the victims of its destructive force has grown even wider. Nowadays the President of the USA or the USSR can authorise the launching of an intercontinental ballistic missile (ICBM) with an atomic warhead that can destroy a large city on the other side of the world. This gap between the attacker and the victim, which has been brought about by technological developments, has two important consequences. First, the attacker does not see directly the human misery that results from the destructive effects of the weapons and so any inhibitions about causing mass destruction of human lives are greatly reduced. Second, the victim cannot indicate the normal gestures of appeasement that are commonplace in non-human animals and in human face-to-face combat. It is no use waving a white flag at an incoming guided missile.

Many governments have devoted a considerable proportion of their scientific research budgets to military research (see Figure 4.1.1). The economic success of Germany and Japan has been attributed to their treaty requirements after World War II, which prevented expenditure on armaments. This allowed them to concentrate funds on research and development of new technology. The need to divert expenditure away from armaments towards more socially useful production has contributed to the Soviet leadership's desire for nuclear disarmament.

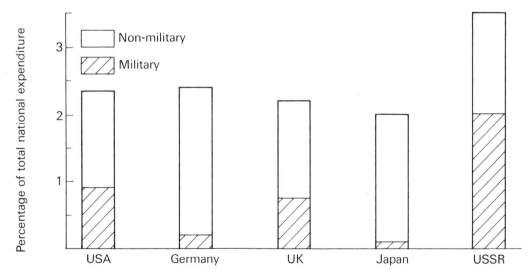

Figure 4.1.1   The percentage of national expenditure on military and non-military research and development (1970–1980).

Science, Technology and Society © D. Andrews, 1992

THE HABER PROCESS AND WORLD WAR I

Until 1908 industry's only source of nitrogenous compounds for manufacturing fertilisers, dyes and explosives was naturally occurring sodium nitrate, mined in South America. Since these supplies were dwindling, scientists were keen to develop a chemical production method. In that year a German chemist, Fritz Haber, made the important breakthrough. He constructed apparatus in his laboratory that could produce the necessary conditions for a reaction between the gases nitrogen and hydrogen to form ammonia. It required a temperature of 600 °C, a pressure 200 times normal atmospheric pressure and a suitable catalyst.

The German military leaders who were preparing for war realised that they would be cut off from vital raw material for making fertilisers and explosives if their enemies enforced a blockade. They were keen to solve the major problem of developing a large-scale production plant that could withstand the high temperatures and pressures required by Haber's process. This could not be achieved without considerable investment.

In 1909 a leading German chemical company, BASF, began to tackle this problem. After four years, during which one million pounds was invested, a huge industrial plant capable of producing 10 000 tonnes of ammonia per year had been designed. The War, which broke out in 1914, would have been over very quickly if the Germans had not achieved this remarkable chemical engineering feat.

LASERS IN WARFARE

In March 1983 the US President, Ronald Reagan, announced a new development in military technology, which he described as the Strategic Defense Initiative (SDI) but which came to be more commonly known as his Star Wars proposal. This initiative called for a massive programme of research and development of new technology to improve the defence capability of the USA against long range nuclear weapons. Some scientists speculated that laser beams fired either from land or from satellites could be used to destroy any incoming enemy missiles. In theory this would create an impenetrable defensive shield, giving the USA complete military superiority. Although still in the experimental stages, this idea greatly worried the USSR who felt it would put them at a disadvantage. They also felt threatened by the possibility that US satellites with laser weapons could be located in orbit above the USSR. However the idea of lasers fired into space as defensive weapons has been ridiculed by some scientists who regard it as technically impossible.

The first application of lasers in warfare occurred in 1991 during bombing raids on Iraq. The missiles fired from planes were guided by following a laser light beam, which had been fixed on the target by the plane's navigator.

ACTIVITY

High-tech war?

1 In 1991 the Gulf War between Iraq and the multinational force led by the USA was described as the 'first high-tech war'. Do you agree that it was? How did the new technology used in this war affect its outcome?

2 Find out in which war the following examples of technology were first used. How did they influence the way that wars are fought?

napalm, spy satellites, computers for code breaking, tanks, radar, 'smart' bombs, 245-T (Agent Orange)

3 Suggest how the invention of gunpowder in the twelfth century might have changed the nature of warfare.

4 In January 1991 an editorial in the New Scientist warned that seeing the horrors of the Gulf War might have a harmful influence on young people's image of science. It said:

> When schoolchildren realise that science and technology are linked with that sort of carnage will they remain enthusiastic about studying either? A career in accountancy is a lot easier on the conscience.

Do you agree that the image of science was harmed by the Gulf War?

4 WAR, TECHNOLOGY AND SOCIETY

UNIT 4.2 The atomic bomb: from Einstein to Hiroshima

At the time a scientific discovery is made it may often appear to have no practical use. It is only with the benefit of hindsight that an important invention can be seen to rely on the previous work of many scientists who may have had no inkling of the potential applications of their discoveries. When these scientists then realise how their work is being applied they may strongly disagree with it and try to stop it. Yet other scientists may take an opposing view, believing that the practical application of the knowledge they have discovered is 'none of their business'. The development of the atomic bomb powerfully illustrates this point. In this unit you will explore the scientific discoveries that were necessary for this devastating new weapon to be developed (see Table 4.2.1). You will also see how the scientists involved tried to influence the political leaders.

It was Albert Einstein who first calculated the tremendous power that could be unleashed from even tiny quantities of matter. His Theory of Relativity, published in 1904, completely revolutionised the theoretical basis of physics. It included the discovery that the mass lost during a chemical reaction would be released as energy according to the following equation:

Energy released (E) = mass lost (m)
　　　　　　　　× velocity of light squared (c^2)

or

$$E = mc^2$$

Since the velocity of light is 300 000 km per second you can see that even a few grams of matter could be converted into vast amounts of energy. At that time there was no immediate prospect of being able to release this energy since it was locked into the nucleus of the atom and could not be tapped by normal chemical reactions. But over the following decades Einstein and many other scientists realised that the time was approaching when the technology to release this energy could be developed. They envisaged two very different forms of nuclear energy release: either very rapidly in the form of an uncontrolled explosion or very slowly under careful control in the form of useful power to generate electricity.

It was the start of World War II in 1939 that pushed research away from peaceful uses and towards the development of the most powerful bomb in the history of warfare. In the USA many scientists believed that Hitler was making rapid progress towards an atomic weapon. Einstein himself wrote to the US President, urging him to back the scientists who were capable of developing an atomic weapon. Many eminent scientists had no hesitation in joining the group working at Los Alamos on the Manhattan Project.

After five years of intensive research effort involving over 3000 scientists, a successful test explosion was carried out on 15 July 1945. A general who witnessed it was awestruck by the sight and sound: 'it ... made us feel that we puny things were blasphemous to dare tamper with the forces heretofore reserved to the Almighty.'

After the test explosion the scientists at Los Alamos began to express their views on the way the bomb should be used more publicly. A popular view was that a demonstration explosion of the atomic bomb should be arranged so that all the world's nations could witness its potential. It was hoped that this would galvanise governments into agreeing a strict code of international control of atomic weapons. However, despite the vigorous opposition from scientists, politicians authorised the dropping of atom bombs on the Japanese cities of Hiroshima and Nagasaki in the first week of August 1945 (see Unit 4.3). The war ended with the unconditional surrender of Japan on 11 August. Although there was much relief that the war was over, there was also grave anxiety about the new era of destructive warfare that the atom bomb had ushered in.

Table 4.2.1   From Einstein to Hiroshima

Date	Scientist	Event
1904	Einstein	Proposes the theory that mass can be converted into energy ($E = mc^2$).
1911	Rutherford	Discovers that the nucleus of the atom is unstable and capable of releasing alpha particles.
1932	Chadwick	Discovers the neutron, one of the particles in the atomic nucleus.
1932	Cockcroft and Walton	Accomplish the splitting of a lithium atom by bombarding it with hydrogen nuclei. Very little energy is released.
1935	Szilard	Suggests that all further experiments involving attempts to bombard atoms should stop because of their potential for releasing massive amounts of energy in an explosion.
1938	Hahn and Straussman	Bombard a uranium atom with neutrons, causing it to split into two, releasing a large amount of energy.
March 1939		News of the German capture of uranium mines reaches nuclear physicists in the USA.
Oct 1939	Szilard and Einstein	Send a letter to the US President Roosevelt advising him to support research into an atomic weapon. They believe that Germany is actively developing such a weapon.
1944	Einstein	Sends a letter to the US President warning him against using the atomic bomb.
1945		Test of the first atomic bomb in the Arizona Desert. Two atomic bombs dropped on Japanese cities.

ACTIVITY

Science and moral choices

1 Before the first test explosion of the atomic bomb one of the leading scientists at Los Alamos, Enrico Fermi, responded to his colleagues' moral qualms about developing an atomic bomb by saying:

> Don't bother me with your conscientious scruples. After all, the thing is superb physics.

This implies that the search for scientific knowledge is divorced from moral questions.

- Why do you think Fermi took this position?
- Do you agree with it?

2 The moral conflict for scientists developing weapons is summarised in these two statements:

- Scientists have a moral duty to ensure that the knowledge they discover is only used for peaceful purposes.
- Scientists must contribute to the development of improved weaponry. It is a necessary evil that must continue as long as the climate of suspicion between hostile nations continues.

What are your views on these statements?

ACTIVITY

Secrecy and science

In peacetime it is standard practice for scientists to share their discoveries by publishing their findings in scientific journals. However scientists involved in military research are required to sign a declaration that they will not disclose publicly any information that might be useful to an enemy. In the UK this is a legal requirement of the Official Secrets Act.

1 How does scientific progress benefit from publication of research results?

2 What is the likely impact of the Official Secrets Act on the progress of scientific knowledge?

3 In the spring of 1945 a physicist working on the atomic bomb, Klaus Fuchs (a German who had become a British citizen in 1933), leaked information to the USSR. He was tried in the UK for high treason and sentenced to life imprisonment in 1950. What do you think motivated Klaus Fuchs to become an 'atomic spy'?

4 WAR, TECHNOLOGY AND SOCIETY

UNIT 4.3 Atomic bomb decisions

After the test explosion of the American atomic bomb on 15 July 1945 at Almogadro in the Arizona Desert (see Unit 4.2), there was heated debate among the scientists involved in the Manhattan Project. The scientists who witnessed it were struck by its frighteningly spectacular and destructive power, but there was disagreement about the best way to use it to bring about an end to the war. In late July some of the scientists decided to take a vote on the future military use of the A bomb. Questionnaires were sent to 150 scientists, who were asked to vote for one of the five options listed in Box 1. There was no opportunity to debate the options but it is likely that the main factors influencing their views were the ones listed in Box 2.

ACTIVITY

How should the atomic bomb be used?

1 How do you think each of the factors listed in Box 2 influenced the way the scientists voted?

2 How would you have voted? Explain your reasons to someone else in the class.

3 Make rough estimates of the percentages of scientists who voted for each of the options in Box 1.

4 Ask your teacher for the actual percentages of scientists who voted for each option. (See *Teacher's Notes*.) How do they compare to your estimates? Are you surprised by the pattern of voting?

5 As a class, conduct a debate on this issue. One person should speak in support of each option for three minutes, and then the rest of the class can ask the speakers questions and discuss the issues raised. Take a vote at the end of the debate.

1 Military options

1 Allow the US military forces to decide how best to deploy the atomic bomb to bring about a rapid end to the war.

2 Give a military demonstration of the bomb in Japan, to be followed by an opportunity to surrender before full use of the weapon.

3 Give a demonstration of the bomb in the US, observed by Japanese representatives, to be followed by an opportunity to surrender before full use of the weapon.

4 Avoid its military use as a weapon but publicly demonstrate its effectiveness.

5 Keep all knowledge of the bomb secret and prohibit its use in the current war.

2 Political influences

1 The Manhattan Project had already cost US taxpayers two billion dollars.

2 In June 1945 US troops were involved in exceptionally bloody fighting to capture control of the island of Okinawa off the coast of Japan. If troops were used to invade Japan it was estimated that 250 000 would be lost.

3 The use of the atomic bomb would demonstrate the military superiority of the US over the USSR and deter the Russians from exerting their influence after the war ended.

4 The huge atomic bomb explosion would cause indiscriminate destruction of civilians and lead to a world-wide reaction of shock, repulsion and condemnation of the US action.

5 The military use of this new type of weapon would lead to an uncontrollable arms race, making future world security more uncertain.

Science, Technology and Society © D. Andrews, 1992

4 WAR, TECHNOLOGY AND SOCIETY

UNIT 4.4 # The medical effects of nuclear war

The explosion of an atomic weapon is very different from a conventional explosion of the same destructive force. As well as the immediate effects of the blast, the explosion produces vast quantities of radioactivity. In this unit you will investigate how people's health was harmed by the effects of radioactive fallout from the nuclear weapons used in Japan.

RADIOACTIVITY AND HEALTH

At 8.15 a.m. on 6 August 1945 a United States B29 bomber plane released an atomic bomb, nicknamed 'Little Boy', over the Japanese city of Hiroshima. The force of the explosion flattened buildings over an area of 20 square kilometres. Within seconds a fireball at a temperature of 3000 °C had spread over the city. One hundred thousand people were killed instantly and over the following months many more people suffered from radiation sickness due to the high dose of radiation they had received. Rapidly dividing cells, such as those in the skin, hair follicles, bone marrow and the lining of the intestines and blood vessels, are particularly vulnerable to the damaging effects of radiation. Radiation sickness as a result of damage to these tissues caused vomiting, loss of hair and internal bleeding. In some cases the impact of radiation damage to the body cells was not apparent for many years, such as a greatly increased risk of leukaemia (see Figure 4.4.1).

When radiation passes through cell nuclei it can damage the DNA molecules and change the genetic information. This can cause the death of the cell, prevent it from dividing or in some cases convert it into a cancerous cell. The cells of the fetus are very susceptible to radiation, which was clearly seen in the aftermath of Hiroshima.

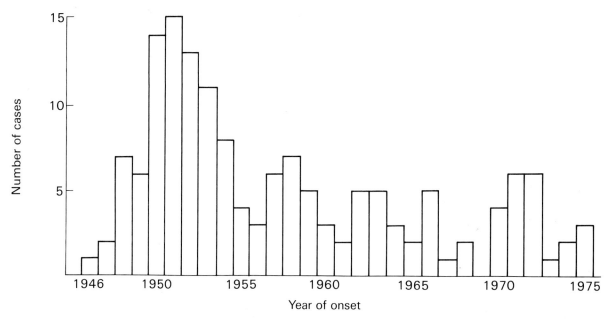

Figure 4.4.1   The number of cases of leukaemia diagnosed each year in Nagasaki from 1945 to 1975.

ACTIVITY

Medical impact of radioactivity on pregnancy

Scientists studied the survivors of the atomic bomb explosions to discover the health effects of the radiation dose they had received. One way of estimating an individual's radiation dose was to determine their distance from the point on the ground above which the bomb exploded. This point is known as the bomb's hypocentre.

Table 4.4.1 summarises the effects of the atomic bomb on the children of different groups of pregnant women who survived. The three groups were defined as follows:

A: women who were within 2 km of the bomb's hypocentre and who suffered from radiation sickness

B: women who were within 2 km of the hypocentre who did not suffer from radiation sickness

C: women who were within 4–5 km of the hypocentre.

Study the data in the table carefully and then answer these questions.

1 Which of the three groups of women received:
   a) the lowest radiation dose?
   b) the highest radiation dose?

2 Calculate the following for each of the three groups:
   a) the percentage of miscarriages
   b) the percentage of mothers whose child died during pregnancy or before it reached 12 months old.

3 Describe the pattern in the data you have calculated in Question 2 and suggest a possible explanation for it.

4 Use the data to investigate the hypothesis that there is a correlation between the size of the radiation dose received by a fetus and its risk of mental handicap.

Table 4.4.1

Group	No. of preg- nancies	Mis- carriages	Still- births	Infant deaths (0–1 yr)	Mentally handi- capped
A	30	3	4	6	4
B	68	1	2	3	1
C	113	2	1	4	0

ACTIVITY

Radioactivity and growth

Table 4.4.2 shows the mean average adult heights attained by men who survived their exposure to radiation at Hiroshima in 1945.

The groups are defined by the men's ages at the time of the bombing and the radiation dose they received.

Study the data in the table and then answer these questions.

1 In comparison to those not exposed, how did very high radiation doses affect the eventual adult height of those in the following age groups at the time of the bombing:
   a) 0–5 year olds   b) 6–11 year olds
   c) 12–17 year olds?

2 Suggest why the same radiation dose did not have the same effect on the growth of the boys in the different age groups.

Table 4.4.2

Age (years)	Not exposed	Radiation dose (rads)		
		0–9	10–99	100+
0–5	166.4	166.1	165.9	161.5
6–11	162.4	164.2	166.3	162.2
12–17	164.3	163.6	164.3	164.4

Science, Technology and Society © D. Andrews, 1992

4 WAR, TECHNOLOGY AND SOCIETY

UNIT 4.5 Controlling the arms race

The explosion of the atomic bomb at Hiroshima in 1945 acted as the starting pistol for the arms race between the USA and the USSR. The production of increasingly advanced nuclear weapons in ever greater numbers was the goal. Other countries tried to join in. The various past attempts and future proposals to control or even reverse the arms race are explored in this unit.

THE END OF THE COLD WAR?

1990 saw the development of much closer co-operation between the USA and the USSR, and the reunification of East and West Germany. So this year was seen by many people to mark the end of the Cold War. International treaties between the USA and the USSR throughout the latter half of the 1980s brought about a considerable reduction in the numbers of soldiers and nuclear weapons based in Europe. The Intermediate Nuclear Forces (INF) Treaty of 1988 resulted in the complete elimination of intermediate and short range missiles by both sides. It also ensured that both sides would have free access to each other's military installations to check that the conditions of the treaty were being carried out.

Despite this welcome breakthrough in nuclear arms control, the nuclear powers still have long range strategic nuclear weapons in submarines and on land. Some people feel that it is morally unacceptable for any country to possess and threaten to use nuclear weapons. British campaigners for unilateral disarmament argue that Britain should take the initiative and be the first country to give up its nuclear weapons voluntarily. This proposal remains a keenly debated issue.

ACTIVITY

The arms control controversy

1 After the USA exploded their nuclear weapon in 1945 they continued to test and develop new bombs. In 1949 the USSR entered the nuclear arms race, followed by Britain in 1955, France in 1960 and China, Israel, India and Pakistan in the 1970s. Suggest why these countries were the first to develop nuclear weapons.

2 The testing of nuclear weapons during the 1950s resulted in a massive increase in the amount of radioactive material released into the atmosphere. This was absorbed into the soil and eventually into cows' milk through the grass they ate (Figure 4.5.1). International concern over this problem led to the Partial Test Ban Treaty in 1963, which outlawed all atmospheric testing of nuclear weapons. What impact did the Treaty have on levels of radioactive fallout as reflected in the level of radioactive strontium in cows' milk?

3 In an attempt to limit the proliferation of nuclear weapons the 1970 Non-Proliferation Treaty was signed by over 60 countries. It prevented the transfer of nuclear weapons technology from nuclear to non-nuclear countries. A number of countries, such as France, China and Brazil, are not signatories to the Treaty.

   a) How effective has the NPT been?

   b) What are the risks attached to nuclear proliferation?

4 Suggest how satellites could be used in the verification of the 1988 INF Treaty.

5 Table 4.5.1 sets out the opposing arguments on the issue of disarmament. Discuss the merits of each pair of arguments.

6 Multilateral disarmament means that opposing countries negotiate an agreement to reduce or eliminate nuclear weapons. The Campaign for Nuclear Disarmament (CND) argues that limited unilateral action could stimulate a multilateral agreement.

> What is required is a country with the intelligence and courage to say, 'Whatever you do, we will make a limited unilateral reduction – and if you respond we will make a further cut-back.'

What are your views on this argument?

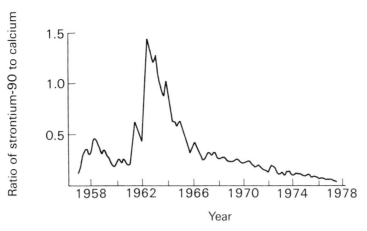

Figure 4.5.1 The average ratio of radioactive strontium-90 to calcium in UK cow's milk (1958–1978).

Table 4.5.1

Do we need nuclear weapons?

Pro nuclear weapons	*Anti nuclear weapons*
A country that possesses submarine-based nuclear weapons could inflict intolerable damage on any other country that launched a nuclear attack against it. This amounts to Mutually Assured Destruction (MAD) and acts as a deterrent to any threatening nation. The fact that there has been no war in Europe since 1945 shows that the principle of deterrence is working and that nuclear weapons are necessary to maintain peace.	The effectiveness of deterrence in the past is no guarantee that it will continue to prevent war in the future. The continued existence of nuclear weapons creates the ever present risk that one may be launched in error. In addition, the deterrent is not credible, since to use nuclear weapons would be to invite immediate retaliation, which would be national suicide.
The NATO powers want to achieve disarmament but they will only negotiate from a position of superior strength in nuclear arms. It is therefore sensible to continue to increase our arms while showing our readiness to negotiate.	If the Soviet government took the same position as this there would be no hope of achieving arms reduction. It is impossible for each side to be the stronger adversary at the same time. 'Arguing from a position of strength' is meaningless when both sides possess 'overkill' – enough warheads to wipe out the entire population of the planet many times over.
What is the point of Britain giving up nuclear weapons if others do not? There is no evidence that this would have any impact on other countries.	It would be a principled act, signalling to the rest of the world our revulsion of nuclear weapons. Our refusal to take part in the arms race would encourage other countries to drop out too.
The USSR cannot be trusted. Some of its generals and politicians see their country as an imperialist power that could dominate the rest of the world. If we give up the bomb what is to stop them from invading us?	The USSR lost 20 million people during Hitler's invasion and have no desire for another war. They would gain no advantage by either invading or wiping out Britain. They believe that their arms are purely defensive against the might of the 'enemy', just as western countries do.
The manufacture of arms creates many jobs. Arms reductions would destroy huge sectors of industry and lead to an unacceptable increase in unemployment.	It would be relatively straightforward for governments to coordinate the conversion of the arms industry to the manufacture of useful products. A lathe can produce parts for a tractor as easily as parts for a tank.

4 WAR, TECHNOLOGY AND SOCIETY

UNIT 4.6   **Chemical weapons**

Poisonous gases were first used in warfare during World War I when many soldiers suffered damaged lungs after breathing mustard gas. Chemical weapons directed at humans have become more sophisticated since then, but they have rarely been used in warfare. Their major drawback as a weapon is that if two opposing armies are in the same area the poison gases can drift back to affect the soldiers who are using them. During the 1980s Iraq's use of chemical weapons in its war against Iran shocked the world.

Chemical weapons again hit the headlines during the Gulf War, which followed Iraq's invasion of Kuwait in August 1990. Soldiers from the international force sent to the area were issued with protective suits in case of chemical weapon attack. In Israel all citizens were issued with gas masks and told to turn one of the rooms in their homes into a sealed retreat. Many scientists raised fears that other developing countries would develop chemical weapons technology. The newspaper article below investigates this problem.

Third World Scramble for Deadly Gases

Third World countries are queuing up to share Iraq's expertise in chemical weapons. Many commentators believe that Iraq's use of deadly gases in its war with Iran has convinced some Third World rulers to begin manufacturing chemical weapons.

Third World countries that seek to expand their military arsenal usually lack the finance and the technological expertise to build nuclear weapons. Chemical weapons, on the other hand, are relatively cheap and simple to make, and the ingredients needed for their manufacture are easy to obtain. Mustard gas, for instance, is made from thiodiglycol, a chemical used in ballpoint pen ink and many other products. Phosphoryl chloride is an ingredient of pesticides, hydraulic fluid and the nerve gas Tabun.

There are four basic categories of chemical weapons.
- Nerve gases
  These chemicals kill within seconds by interfering with the function of the nervous system. Iraq used the nerve gas Tabun in its war against Iran.
- Blood gases
  These lethal gases, such as hydrogen cyanide, are carried in the bloodstream, enter cells and block the activity of vital enzymes.
- Damaging agents
  These gases damage any surface with which they come into contact. Mustard gas, for example, burns the skin and damages the lungs.
- Choking agents
  When these gases are inhaled, they cause the lungs to collapse. Phosgene was the poison gas most widely used in World War I.

The USA and the USSR hold the largest stockpiles of chemical weapons. Their combined stocks amount to 80 000 tons. Britain destroyed all its stocks in the 1950s and the USSR announced in 1988 that it would begin unilateral dismantling of its own chemical weapons. At an international conference in Paris, 142 countries voted unanimously to renounce the use of chemical weapons. However, several Third World countries insisted that they could not be expected to give up chemical weapons unless neighbouring hostile states agreed to remove the threat of nuclear weapons. The uncertainty of exactly which countries have chemical weapons adds to the impetus for proliferation.

The 1991 negotiations in Geneva to formulate an international treaty to ban chemical weapons became bogged down over the question of verification. Several recent events have shown that it is not easy to distinguish between factories manufacturing pharmaceutical products and those that produce lethal chemicals for use in warfare.

ACTIVITY

Political, military and technological issues

Read the newspaper article and then answer these questions.

1. Give two reasons why Third World countries are able to develop chemical weapons but not nuclear weapons.

2. Why are the superpowers in the West more willing to give up chemical weapons than are some developing countries?

3. What is meant by the 'impetus for proliferation', and what is the main factor causing it?

4. What is meant by 'verification' and why does it play such an important role in the control of proliferation?

5. Suggest a strategy that an independent team of scientists could use to decide whether a plant was being used for the manufacture of chemical weapons.

6. Find out more about the biochemistry of the chemical substances mentioned in the article. For example, how exactly do Tabun and hydrogen cyanide disrupt normal cell function?

7. Using Figure 4.6.1, explain why the manufacture of mustard gas does not require highly advanced technology.

8. It has been suggested that suppliers of phosphoryl chloride should report any significant losses or thefts so that the authorities are alerted to the possible unauthorised manufacture of chemical weapons by terrorist groups. Do you believe this tactic would be:
   a) practical to implement?
   b) successful in its aim?

9. If you were a scientist, would you take a job that required you to develop chemical weapons?

10. Why do you think some scientists have been prepared to work in this field of research?

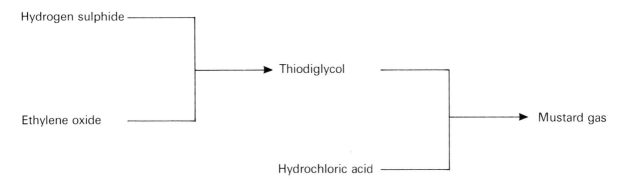

Figure 4.6.1 The chemical pathway for the manufacture of mustard gas.

Science, Technology and Society © D. Andrews, 1992

5 THE REPRODUCTION REVOLUTION

UNIT 5.1 Natural selection at work

Scientists aim to develop explanations for the observations that they make. But the same set of observations may have several different possible explanations. Charles Darwin's theory of natural selection is an excellent example of how one explanation came to be accepted in preference to several others.

In Darwin's time a number of alternative theories had been put forward to explain the fossil evidence of evolution (see below). He found all of them unsatisfactory and so he began to develop his own theory to explain how evolution had occurred. His ideas were stimulated by his observations of the diversity of plant and animal species he encountered during a five-year round-the-world voyage on the HMS *Beagle* (1831–6). The most famous of these is his intensive study of the many different species of finch that populated the islands of the Galapagos Archipelago in the South Pacific. A comparison of these species suggested that they had all evolved from a common ancestor and provided very strong evidence to support Darwin's theory of natural selection. Figure 5.1.1 shows how the shape of the finch's bill is related to its diet. One species has evolved a sturdy bill for crushing seeds; another has evolved a longer bill for probing for insects in tree bark.

Finch species with a bill adapted for crushing large seeds

Finch species with a bill adapted for probing tree bark for insects

Figure 5.1.1   Two types of finch found in the Galapagos Archipelago.

EARLY THEORIES OF ORIGINS

Lamarck

The French biologist, Jean Baptiste de Lamarck (1744–1829), proposed that simpler species change over the generations into more advanced species. These changes were seen as an inevitable process triggered by the organisms' efforts to overcome the challenges posed by the environment. Characteristics acquired in this way during an individual's lifetime can be passed on to its offspring. According to Lamarck's theory, the giraffe's long neck develops through many generations of trying to stretch up to reach the highest tree branches.

Catastrophism

According to this view, there has been a series of global catastrophes at various points in the Earth's history, which have wiped out all the living species on the Earth. After each of these mass extinctions, the last of which was the Great Flood as recorded in the Bible, God miraculously created a new set of more advanced life forms.

THE PROCESS OF NATURAL SELECTION

To summarise Darwin's theory of natural selection, organisms are capable of producing large numbers of offspring. However, many of these offspring die before they reach adulthood and before they can reproduce. They die because they are unable to overcome many difficulties: these include the dangers of being eaten by predators, the risks of suffering disease or injury and the problems of finding enough food. Darwin said that there was a **struggle for survival**. However, he made it clear that the struggle was not necessarily a conflict between individuals. Organisms also struggle against the life-threatening demands of a harsh environment, such as extreme heat or cold. Scientists now describe these difficulties as **selection pressures**.

As there is variation between individuals of the same species, some are better equipped than others to overcome these pressures. For example, the antelope with the most powerful leg muscles will be the one that is most likely to escape from a fast-moving predator, such as a cheetah. Features like this, which help the individual to survive, are known as **adaptations**. Darwin observed that many adaptations are passed down from parents to their offspring. We now know that the information for adaptations is carried in the form of genes.

Darwin argued that organisms that survived to overcome the difficulties of the environment would leave more offspring, which would inherit their parents' adaptations. Organisms that were not so well adapted would leave fewer or no offspring. Therefore, after many generations the population would consist of organisms better adapted to the environment. If the environment changes, natural selection chooses those genetic variations that give an advantage in the new conditions. Random changes in the chemical structure of genes, otherwise known as **mutations**, continually produce new genetic variation. If such a mutation is advantageous to the organism that inherits it, it will be the basis for an improved adaptation and spread through the population with each generation. Selection will therefore encourage the change in form that we call **evolution**.

ACTIVITY

Natural selection at work

In parts of Africa and Asia a mutant gene, which is normally harmful, instead gives the individual an advantage in the struggle for survival, and so provides a good example of natural selection operating in a human population.

The gene that causes sickle cell anaemia, a serious inherited disorder, is more commonly found in regions where there is a high incidence of malaria, an infectious and often fatal disease (see Unit 5.3). People with two sickle cell genes have a very high proportion of sickle-shaped red blood cells. In developing countries these people usually die before having children. We would therefore expect this gene to be removed from the population by natural selection. The fact that it is surprisingly common led scientists to suggest that people with one sickle cell gene and one normal gene might be protected against malaria. Such individuals have a certain proportion of sickled red blood cells but do not suffer from sickle cell anaemia.

1 Compare the two maps in Figure 5.1.2.
   • Explain any patterns you observe.
   • Explain why this is indirect evidence of selection at work.

2 Examine the data in Table 5.1.1, which summarises the results of blood tests on 290 people living in a malarial zone. What conclusions can you draw from it?

Table 5.1.1

	Malaria parasites in blood	No malaria parasites
People with sickle cells	12 (27.9%)	31 (72.1%)
People with no sickle cells	113 (45.7%)	134 (53.3%)

(a)

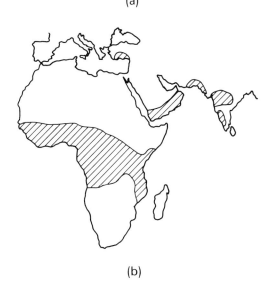

(b)

Figure 5.1.2 The regions of Africa (a) where malaria is endemic; (b) where there is a high incidence of sickle cell anaemia.

Science, Technology and Society © D. Andrews, 1992

5 THE REPRODUCTION REVOLUTION

UNIT 5.2 **Basic genetics**

How are characteristics passed from one generation to the next? This is the fundamental question that the science of genetics seeks to answer. The basic principles of genetics were discovered over 140 years ago by a monk, Gregor Mendel.

THEORIES OF INHERITANCE

Charles Darwin's theory of natural selection, when it was published in 1856, was incomplete without a theory that explained how characteristics could remain distinct as they were passed down the generations. Darwin died ignorant that such a theory had been developed by a monk, Gregor Mendel, working in his monastery garden in the Austrian Alps.

Mendel's new theory of inherited 'factors' that remain distinct as they pass from one generation to the next made him world famous after his death. Although the scientific discoveries he published in 1866 revolutionised our understanding of inheritance, during his lifetime no one recognised the importance of his work and he died in obscurity. The science of genetics failed to make any progress until his ideas were rediscovered in 1900.

Mendel developed his theory of inheritance during seven years of painstaking breeding experiments using pea plants. He concentrated on characteristics of the plants that allowed them to be classified into two distinct groups: for example, whether their pods contained smooth or wrinkled peas or whether their stems were tall or short. Repeating his experiments with seven different pairs of characters gave identical results. Figure 5.2.1 shows the results of one set of Mendel's breeding experiments and how he used the idea of inherited factors – genes – to explain them.

Mendel realised that the inherited characteristics he was investigating were controlled by factors (genes) operating in pairs in all the body cells. In contrast, the gametes (sperm and eggs) that passed on the vital information to the next generation had only one copy of a particular gene. We now know that the gametes are formed from body cells through a special type of cell division called meiosis. Meiosis ensures that each gene in the pair is separated into two different gametes.

So, each parent contributes one gene from its pair through its gametes to its offspring. When Mendel crossed the short-stemmed with the tall-stemmed plants, the offspring received one short stem gene and one tall stem gene. The dominant gene (in this example the tall stem gene) is expressed in preference to the recessive gene (short stem gene), when they occur together in the first generation (the F_1 generation). When the F_1 generation of tall-stemmed plants were crossed with themselves, they produced both tall- and short-stemmed offspring in a ratio of approximately 3 : 1. Mendel explained this by considering the four equally likely combinations of genes produced by the F_1 plants. Only one of the four combinations brought together two recessive genes and produced a short-stemmed plant.

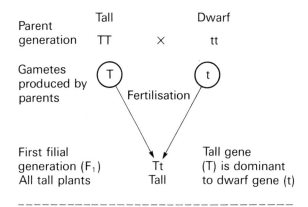

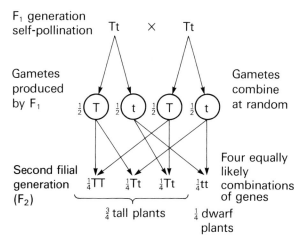

Figure 5.2.1

CHROMOSOMES AND GENES

In the early twentieth century scientists began to build on Mendel's work and knowledge of genetics expanded rapidly. Genes were seen to be arranged along the length of the chromosomes – the thread-like structures found in the nucleus of every cell. The largest chromosomes contain 10 000 genes. During cell division the chromosomes shorten and thicken and can be stained so that they become visible under the microscope. When magnified and photographed they can be seen to form two sets. In modern times, the study of chromosomes in humans has been of great value in diagnosing certain inherited abnormalities such as Down's Syndrome (see Unit 5.5).

DNA, GENES, PROTEINS AND CHROMOSOMES

In 1953 one of the most important scientific breakthroughs of the twentieth century was made. The detailed chemical structure of the gene was discovered after many decades of speculation.

The DNA (deoxyribonucleic acid) molecule was found to consist of a double strand linked together by simple units called bases. There are four types of base: A, T, G and C. A will only pair up with T, and C and G always pair with each other. Figure 5.2.2 shows how this specific pairing allows DNA to copy itself.

A gene contains the instructions for building one type of protein. The DNA bases work in groups of three to build up a chain of amino acids. To do this a 'mirror image' copy of the gene's bases is made, called the messenger RNA molecule. It moves out of the nucleus into the cytoplasm of the cell. A particular triplet of bases attracts one of 20 different amino acids, each attached to its own specific molecule of transfer RNA (tRNA). For example, the base triplet TCA is the code for the amino acid serine. The triplet codes are read off until a complete protein chain of several hundred amino acids is produced (see Figure 5.2.3). A change in one of the bases will produce a change in the amino acid of the protein. This can drastically change the shape of the protein and seriously affect its function. (See Unit 5.3, sickle cell anaemia.)

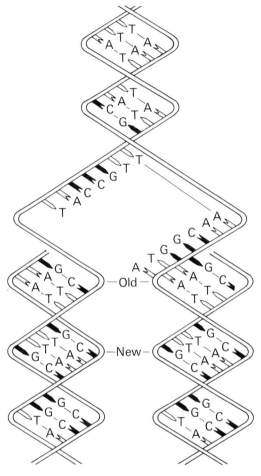

Figure 5.2.2   The replication of DNA.

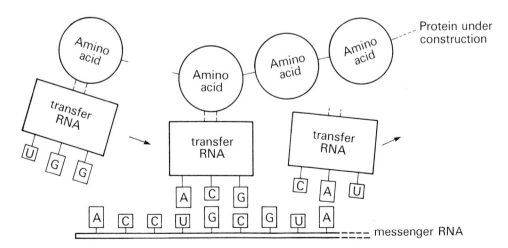

Figure 5.2.3   Building up a chain of amino acids.

Science, Technology and Society © D. Andrews, 1992

ACTIVITY

Breeding by coin tossing

As you have seen from Mendel's investigations, carrying out breeding experiments on plants and animals is often a lengthy process. In humans it is, of course, impossible. However, in this activity you can perform a simulation of a breeding experiment in less than 30 minutes. By tossing coins you can simulate the breeding of a family with as many as 100 children!

a) Take two coins and attach a small sticky label to each side. On one side of each coin write the capital letter 'N', representing the normal dominant gene. On the other side of each coin write the lower case letter 'n', representing the recessive gene responsible for causing an inherited disorder such as cystic fibrosis (see Unit 5.4).

b) Having decided which coin represents the ovum and which the sperm, toss them and record the combination of letters that fall uppermost in a table (see Table 5.2.1).

Table 5.2.1

Egg	Sperm	Number of offspring	Total
N	N	⊦⊦⊦⊦ I	
n	N	I I I I	
N	n	⊦⊦⊦⊦ I I	
n	n	⊦⊦⊦⊦	

c) Repeat the coin tossing until you have at least 60 offspring. Add up the totals for each combination.

Now consider the following questions:

1 How many of the offspring:
   a) are normal   b) are affected by cystic fibrosis?

2 Work out the ratio of normal to affected offspring for:
   a) your own results
   b) the pooled results for the whole class.

3 Compare the answers to 2a) and 2b). Use this comparison to suggest an explanation for geneticists' preference for having large numbers of offspring in their breeding experiments.

4 Why is coin tossing a suitable way of simulating meiosis and fertilisation?

WATSON, CRICK AND FRANKLIN AND THE RACE FOR THE DOUBLE HELIX

After World War II many scientists looked on molecular biology as an exciting area for research. The most important molecule to be found in living organisms was still a mystery – the molecule that stores the information about inherited characteristics.

In the late 1940s deoxyribonucleic acid (DNA) was identified as a prime candidate and from this point there was a race on to work out its exact structure. Linus Pauling, a very eminent US scientist, was the first to publish a possible structure of the DNA molecule, consisting of three intertwined chains. At Cambridge University a young American biologist, James Watson, did not believe that Pauling's structure was correct. As a biologist Watson knew that anything to do with sex comes in twos and he had a hunch that DNA's involvement in sexual reproduction must mean that a double chain was a more likely structure.

Watson teamed up with Francis Crick, another young scientist, and they began to try out their own versions of the DNA structure using cardboard cut-outs to represent the shapes of the different bases. After many months of work they published an article in March 1953 that accurately described the double helix structure of DNA.

Watson and Crick's discovery would not have been achieved without access to the work of another scientist. Rosalind Franklin was a dedicated and methodical researcher working at the University of London's King's College. She spent long hours in the laboratory perfecting a technique called X-ray crystallography, which allowed her to obtain images of the DNA molecule (see Figure 5.2.4). From these she calculated the exact dimensions of the DNA molecule, which were then used by Watson and Crick to build their model. Until recently, Franklin's major contribution has not always received proper acknowledgement in accounts of the discovery of DNA. This story shows that it takes more than one approach to research to achieve a scientific breakthrough.

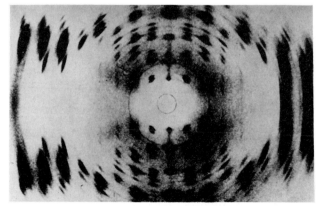

Figure 5.2.4   An image of a DNA molecule created using X-ray crystallography. Scientists can use such an image to calculate the exact dimensions of the molecule.

5 THE REPRODUCTION REVOLUTION

UNIT 5.3 Harmful genes 1: sickle cell anaemia

Some diseases are found more often in certain families than in others. They are inherited or passed on in particular patterns from one generation to the next by means of genes. Each of us carries at least four different harmful genes but, fortunately, they do not show up in either ourselves or our children. This is because most of the harmful genes are recessive. Usually the recessive gene is paired with a dominant gene, which masks its influence so it does not affect our health. However, we can pass it on to our children. If a child receives two similar, harmful recessive genes, one from each parent, he or she will suffer from a genetic disorder (see Unit 5.2). Scientists have identified about 4000 different harmful genes. In this unit you will learn about one of the most common of these: sickle cell anaemia.

Sickle cell anaemia is a genetic disorder caused by a recessive gene that alters the shape of the protein, haemoglobin, found inside red blood cells. Haemoglobin binds to oxygen as it enters the lungs and carries it around the bloodstream to the cells. People with sickle cell anaemia have haemoglobin that is a very inefficient carrier of oxygen and so they easily become breathless. Their red blood cells take on a sickle or crescent shape when their oxygen supply is low (Figure 5.3.1). When this happens inside capillaries, the 'sickle cells' stick together, blocking the blood flow. This causes severe pain in the joints and abdomen.

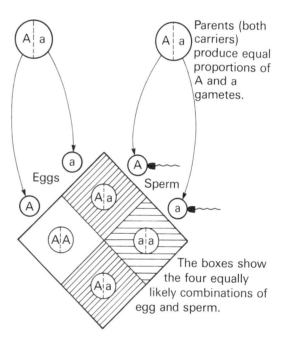

A = Normal dominant gene
a = Recessive sickle cell gene

Parents (both carriers) produce equal proportions of A and a gametes.

The boxes show the four equally likely combinations of egg and sperm.

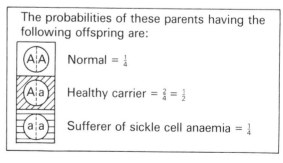

The probabilities of these parents having the following offspring are:

Normal = $\frac{1}{4}$

Healthy carrier = $\frac{2}{4}$ = $\frac{1}{2}$

Sufferer of sickle cell anaemia = $\frac{1}{4}$

Figure 5.3.2 The inheritance of sickle cell anaemia.

Figure 5.3.1 Micrograph of a normal red blood cell and sickle cell red blood cells.

Science, Technology and Society © D. Andrews, 1992

People who have inherited two recessive sickle cell genes will suffer these symptoms. Those who inherit one recessive and one normal dominant gene are said to have sickle cell trait. This means that they will have no symptoms, but since they are 'carriers' of the sickle cell gene they could pass on the disorder to their children. Figure 5.3.2 shows how the chance of this happening can be predicted for the children of parents who both have sickle cell trait.

Sickle cell anaemia is particularly common in central Africa and in people of African origin living in other parts of the world. There is also a low incidence of it in Asian populations. Scientists have found that people with sickle cell trait gain protection against the common tropical disease, malaria (see Unit 5.1). They survive longer and have more children than those who do not have the sickle cell gene. In Britain about one in 400 people of African and Afro-Caribbean origin suffer from sickle cell anaemia and about one in ten are carriers. The passing down of harmful genes from parents to children can be shown on a family tree like the one in Figure 5.3.3.

ACTIVITY

Tracking the gene for sickle cells

You can see from the family tree in Figure 5.3.3 that Alison has sickle cell anaemia. This means that she has inherited two recessive genes. Each of her parents has one dominant normal gene and one recessive gene, like the parents in Figure 5.3.2. Use your understanding of genetics to track the way that the sickle cell is passed down so that you can answer these questions about the individuals shown on the family tree.

1 Give a reasoned explanation for each of the following:
   a) Although Beth has sickle cell anaemia none of her children have.
   b) Colin and Diane are both healthy but their son Ethan has sickle cell anaemia.
   c) After blood tests Diane is found to be a carrier but her brother, Frederick, is not.

2 Colin and Diane would like to have another baby. A genetic counsellor has explained to them that their risk of having a child with sickle cell anaemia is one in four. They believe that since they already have one child who suffers from sickle cell anaemia, their fourth child must be healthy. What is wrong with this argument?

3 Suppose Diane becomes pregnant and is offered a test to detect whether the fetus will suffer from sickle cell anaemia. List the advantages and disadvantages of accepting the offer.

4 Role play a discussion between Colin and Diane. Diane is trying to make up her mind about having the test.

5 Imagine you were in a similar situation. What would you choose to do?

6 You can find out more about sickle cell anaemia by writing to:
Organisation for Sickle Cell Anaemia Research (OSCAR), Cambridge House, 109 Mayes Road, Wood Green, London N22

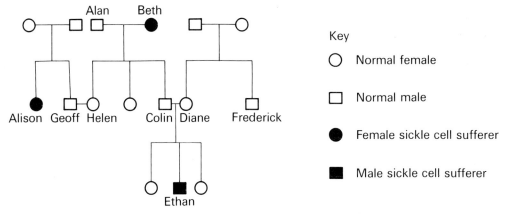

Figure 5.3.3   Family tree showing the inheritance of the sickle cell gene.

SCIENCE TECHNOLOGY AND SOCIETY

5 THE REPRODUCTION REVOLUTION

UNIT 5.4 Harmful genes 2: cystic fibrosis

Cystic fibrosis (CF) is the most common genetic disorder among white people but very rare among non-whites. It is caused by a recessive gene and it is inherited in the same way as sickle cell anaemia. At present, CF cannot be cured and it is rare for a CF sufferer to live beyond 30 years. Until recently, CF sufferers' main hope for a successful treatment has been a combined heart and lung transplant. However there is growing optimism that a new therapy may be developed as a result of a major discovery made in August 1989.

After six years of research, two teams of scientists from different laboratories in Toronto, Canada and Washington, USA discovered the exact location and structure of the CF gene. This led to the development of improved screening tests to detect people who carry a defective copy of the gene. It also raised hopes of better drug treatments or even gene therapy to replace the defective gene itself.

In August 1990, scientists using genetic engineering techniques brought the reality of gene therapy for CF a little closer (see Unit 6.5). They demonstrated that it was possible to introduce the normal version of the CF gene into lining cells of the lungs of live mice. Although this is a promising breakthrough, it is very difficult to estimate how long it will be before it is safe to use this treatment on human CF patients.

ACTIVITY

Choices for CF carriers

The article by Sharon Kingman appeared in the newspaper the *Independent on Sunday* in March 1990. It describes some of the difficult choices that are raised by the new tests, which can detect carriers of the CF gene and affected fetuses. Read this article before answering the following questions on your own. Then discuss your answers with a partner.

1 Explain why CF sufferers need to:
   a) monitor their diet carefully
   b) have regular physiotherapy.

2 Would you want to find out if you were a carrier of a harmful gene? Give your reasons.

3 What would be the advantages and disadvantages of having this knowledge?

4 It has been suggested that all adults should be tested to discover whether they are carriers of cystic fibrosis. What do you think about a compulsory mass screening scheme like this?

5 The article gives three possible options for couples who discover that they are both carriers.
   - What are these options?
   - Which one do you think you would choose if you were in this position?

6 It has been argued that genetic screening of the population could lead to complete government control over the rights of the individual to reproduce. Discuss whether you think this is likely to happen.

7 If doctors are not completely sure of the safety of the newly developed gene therapy, do you think there will be many CF sufferers willing to take the risk of being a subject in the early experimental human trials? Give reasons for your answer.

8 To find out more about CF write to: Cystic Fibrosis Research Trust, Alexandra House, 5 Blyth Road, Bromley, Kent, BR1 3RS

Cruel genes make hard choices

Sharon Kingman on a new test for cystic fibrosis

ONE YEAR when Janet Jacques was collecting money for the Cystic Fibrosis Research Trust, a woman donor said sympathetically: "Oh, cystitis is such a terrible thing!" She did not realise her mistake: she was thinking of the unpleasant but not so dangerous urinary problem. Cystic fibrosis is much more serious: it is a life-threatening disease that affects one in every 2,000 babies born in Britain.

People are ignorant of cystic fibrosis, but they are unlikely to remain so for long. Within the next few years, couples who are thinking about babies will be able to have a simple test to determine the risk of having a child with cystic fibrosis. It is the commonest genetically-inherited disease in Britain.

One in 20 of the British population – two million people – are carriers. They have no symptoms of the disease. The chance of two carriers becoming partners is one in 400, and if they have a child together, there is a one-in-four chance that it will have cystic fibrosis.

The disease makes the body produce unusually thick and sticky mucus. This clogs up the lungs and leaves the child prone to infections. It also blocks the duct from the pancreas, preventing vital enzymes from reaching the gut and digesting food.

Children with the disease have to undergo at least two half-hour sessions of physiotherapy every day (more if they have a cold), to help drain the chest of sticky secretions. They also take many antibiotics to keep infections at bay. If the pancreas is affected, they must take enzyme supplements to help them digest food. Twenty years ago, few children with cystic fibrosis survived into their teens. Now, the average age of survival is 28 for men, 23 for women and increasing all the time.

Last August, medical researchers in Toronto, Canada, announced that they had found the gene responsible for most cases. This means it is now possible to identify most people who carry the defective gene, using a sample of blood or cells from the lining of the mouth.

Several laboratories in Britain are already offering the test to immediate relatives of families who have had a child with cystic fibrosis. Some of Mrs Jacques's family have had the test. Her son, Adam, 11, has cystic fibrosis. His sister, Claire, 14, was diagnosed as a carrier some time ago. Mrs Jacques encouraged her brother and sister-in-law to go for testing. Luckily, neither of them turned out to be carriers. Her sister-in-law, Kim Hirschfield, who is not planning to have children yet, says: "It relieved us of having to face a stressful situation in the future."

What is the choice for couples who are both carriers? Some will choose not to have children. Others will try to adopt or have artificial insemination. For couples who decide to go ahead with pregnancy, prenatal diagnosis is available, with the option of termination if the foetus is found to be affected.

Mrs Jacques says: "If there's a genetic illness in the family, you don't want to see your child suffer." She does not agree with abortion but "until you're in the situation yourself, I don't think you can pass comment".

A mother who does not want to be named, living in London, has a daughter aged 17 who has the disease. She says that if she had had a test and the foetus had proved to be affected, "I think I probably would have had the pregnancy terminated in view of what I know now about cystic fibrosis". Although her daughter's condition is relatively mild, the woman says she would not knowingly want to bring another baby into the world now that she understood how he or she could suffer. Her son, aged 20, has not so far been tested but "I think he will want to know when it comes to having children".

Even families who would not terminate a pregnancy because of their religious beliefs say they would still like to have the information. Mave Salter, whose daughter Claire, 14, has the disease, says: "I would like to know if the baby was going to have cystic fibrosis or not. I would not agree to an abortion. But I would rather know early on and get used to the idea." Claire agrees. "If I had been terminated, it would have been a waste of a life." But until the child is born, it is impossible to predict how severe the disease will be.

Elizabeth Lobo, who is 25 and also has cystic fibrosis, thinks it would have been better if her mother had had prenatal diagnosis and an abortion. "I don't regret being born, but I think screening is very advantageous and could wipe out cystic fibrosis, which would be a good thing."

Dr Maurice Super, consultant paediatric geneticist at Royal Manchester Children's Hospital, says there is strong evidence that families that have one child with cystic fibrosis tend not to have any more. He has kept a register of women who have had prenatal diagnosis: most have opted for termination on discovering the foetus is affected.

© *The Independent on Sunday*, March 1990.

5 THE REPRODUCTION REVOLUTION

UNIT 5.5 Amniocentesis and Down's Syndrome

All parents are anxious about their child as it develops during the nine months in the womb. Could something have happened to interfere with its normal development? Will it be born with a handicap? One way of checking this is to investigate the fetus's chromosomes. The chromosomes with their instructions for constructing the cells have been passed down in the egg and sperm of the parents (see Unit 5.2).

When chromosomes are copied there is always the possibility of a mistake being made. These mistakes, known as **mutations**, can affect either a single gene, as in cystic fibrosis (see Unit 5.4) or a whole chromosome as in Down's Syndrome. Mutations are not always harmful. Some will have no effect at all (neutral mutations) while others, such as the gene responsible for sickle cell, may even give the organism an advantage (see Unit 5.1).

Recent advantages in technology have given us, for the first time, the opportunity to detect harmful chromosome mutations. This unit examines one example of this technology and the ethical problems that it raises.

AMNIOCENTESIS

Since the 1960s, a medical procedure called **amniocentesis** has allowed doctors to examine the chromosomes of the fetus to check for any mutations.

During pregnancy the fetus develops inside the uterus, enclosed within a fluid-filled membrane called the amniotic sac. Skin cells from the fetus are shed into the amniotic fluid. At around the eighteenth week of the pregnancy a sample of amniotic fluid containing fetal skin cells can be withdrawn by a doctor using a specially designed hypodermic needle and guided by an image from an ultrasound scan (see Figure 5.5.1).

The cells are transferred to a dish where they multiply in a nutrient medium. After about five days a chemical (colchinine) is added that interferes with cell division so that the chromosomes are 'frozen' at the stage where they are shorter and thicker than normal. A microphotograph of the stained chromosomes from one of the cells is taken. Figure 5.5.2 shows what this microphotograph could look like.

The chromosomes are cut out from the photograph and arranged in pairs according to size. This is the person's **karyotype**. Figure 5.5.3 shows the karyotype of a normal male. The whole process, from taking the sample to obtaining the karyotype, takes about 10 to 14 days.

There is a slightly increased risk of a miscarriage after an amniocentesis, so it is not recommended unless there is a relatively high risk of an abnormal fetus. If an abnormality is detected, the pregnant woman is offered an abortion by hormone induction. This involves giving a hormone injection that causes muscle contractions in the uterus as in a normal labour. Unlike the suction method used in earlier abortions that can be performed in five minutes, this procedure may take several hours before the fetus is eventually expelled.

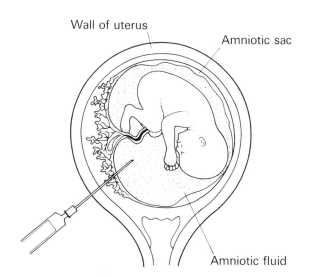

Figure 5.5.1 Amniocentesis is used to collect a sample of amniotic fluid.

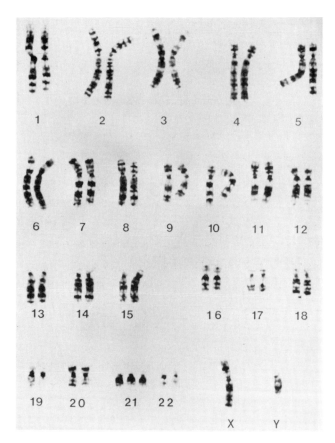

Figure 5.5.2   Microphotograph of a set of chromosomes.

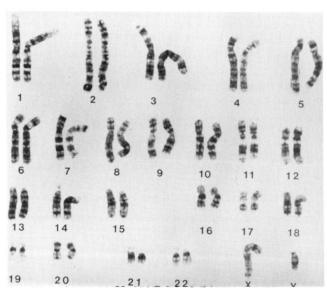

Figure 5.5.3   Karyotype of a normal male.

ACTIVITY

Karyotyping

Cut out the chromosomes shown in Figure 5.5.2. Match up the homologous pairs of chromosomes and arrange them in sequence as in the karyotype shown in Figure 5.5.3. What is the major abnormality that you have identified?

DOWN'S SYNDROME

Normally the gametes have half the number of chromosomes found in the other body cells. The special type of cell division that produces them (meiosis) ensures that the homologous chromosomes are separated. As a result only one from each pair goes into each gamete. The pairs are re-established when the two sets of chromosomes come together in the fertilised egg.

In rare cases meiosis fails to separate one of the pairs so both of the homologous chromosomes end up in the same gamete cell. This can produce an egg with either one extra or one fewer chromosome. If this egg is fertilised, the resulting embryo will be abnormal in some way. In many cases the imbalance in the normal chromosome number will cause such severe disruption to the development of the fetus that it will not survive for the length of the pregnancy.

In some cases the fetus does survive but is born disabled. The most common example of this is when there is an extra chromosome 21, a condition described as trisomy 21. This produces a distinctive pattern of physical and mental characteristics first identified by Dr John Down in 1866. People with Down's Syndrome have distinctive facial characteristics such as a flattened face and large upper eyelids. They also suffer congenital heart abnormalities and have a mental age often as low as six years and unlikely to progress beyond ten years. Down's children are often very affectionate, trusting and friendly and with intensive training they can develop many useful skills. However there is a wide variability in the severity of the mental retardation and this makes the decision to abort a fetus with trisomy 21 a very hard one.

ACTIVITY

Your views on amniocentesis

1 Study Figure 5.5.4 and explain why an amniocentesis is strongly recommended for pregnant women over 35, but is rarely given to pregnant women younger than this.

2 An alternative to amniocentesis is chorionic villus sampling (CVS). A sample of the part of the placenta formed from fetal cells – the chorionic villi – is removed and tested. The test can detect chromosome abnormalities as early as eight weeks after fertilisation. Suggest why this method might be preferable to amniocentesis.

3 If you were offered an amniocentesis would you accept the offer? Explain your reasons.

4 If the results of the amniocentesis showed Down's Syndrome, would an abortion be acceptable for you? Give your reasons.

5 If more money and staff were available to provide support for the parents of Down's Syndrome children, how would people's attitude to Question 4 change?

6 Some doctors only offer to carry out an amniocentesis if the woman agrees in advance to have an abortion should the fetus show any abnormality.
  - Suggest why these doctors take this view.
  - Do you believe they are justified?

7 Suggest why a 38-year-old pregnant woman might refuse the offer of an amniocentesis. If you were a doctor faced with this reaction what would your attitude be?

8 Down's Syndrome was previously called mongolism because it was thought that the characteristic facial features were similar to those of Mongolian people. Suggest why this name for the condition is no longer used.

ACTIVITY

Role play

You can explore these issues further by staging a series of role plays in groups of three. One person takes the role of the doctor, another the role of a 38-year-old pregnant woman and the third the father of the child.

To prepare for the role play, write down some key background facts on the three roles, e.g. their names, religious beliefs, views on abortion, family situation.

Scene 1
The couple are discussing with the doctor whether to have an amniocentesis, and what they will do if the result shows that the fetus has trisomy 21.

Scene 2
The couple have just been told the result of the amniocentesis. The fetus has trisomy 21. They discuss what to do next.

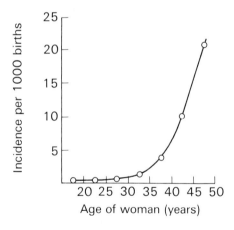

Figure 5.5.4   A woman's chances of having a baby with Down's Syndrome at different ages.

5 THE REPRODUCTION REVOLUTION

UNIT 5.6 Images of disability

What are the common images of disabled people? Are they widely believed to be severely limited in their capabilities? The weekly BBC Radio 4 programme on issues concerning the disabled is titled *Does He Take Sugar?* It was chosen because this question had been put to a parent accompanying her wheelchair-bound son. Of course, he was of normal intelligence and perfectly capable of answering for himself. What assumptions had been made about the physically disabled by the person who asked the question?

Figure 5.6.1

ACTIVITY

Behind the myth

The newspaper article should make you rethink some of the popular myths about disability. It describes the rock group, Heart 'n' Soul, whose members are all either physically or mentally disabled. Read the article before discussing these questions.

1. Has the article changed your image of people with Down's Syndrome? In what way has it changed?

2. How do you think that other people's negative attitudes to disability affect the lives of disabled people?

3. The members of Heart 'n' Soul devised a show based on their own lives that was 'about a group of people who have been loved too much'.
   a) Why do you think they called the show 'The Dungeon Of Love'?
   b) Suggest reasons why the carers of disabled people might be likely to devote themselves too much to their task.
   c) What are the likely effects on disabled people of being 'loved too much'?

4. Choose a particular musical instrument and a specific type of disability. Suggest how the musical instrument could be adapted to make it easier for a person with that disability.

5. Find out more about the difficulties of people who are physically or mentally disabled by building contacts with one of the following pressure groups in your local area. You might also wish to invite a disabled person to come in to talk to your class.
   MENCAP
   Down's Syndrome Association
   Spastics Society

Figure 5.6.2   Christopher Goodall, from the rock group 'Heart 'n' Soul'.

Heart at the heart of the matter

John Cunningham

SOMETIMES one or two people come in when they are drunk and start taking the mickey, Christopher Goodall says. "Once I had a dizzy spell..." He does not need to add anything: lead singers with any group have to take the flak sometimes.

From their music, there is nothing to tell you that no one in Heart 'n' Soul looks anything like a rock performer. For all its members are severely disabled, either mentally or physically. Goodall, for instance, has Down's Syndrome, and all the songwriters, singers and musicians have faced life in day centres until music thrust them into the spotlight.

Or, rather, they thrust themselves there. Goodall, 28, has the ego of Gazza and the sparkiness of Jonathan Ross. It is impossible to imagine his disability bottling up his thirst for showmanship. School plays, intended as therapy, started him off; as music classes did for other members of the group of performers who are probably unique.

Their sound is a mixture of old pop, reggae, ballads and funk. They have made a video and a new single is released next week. The venues, in the five years since they have been together, have mostly been clubs and conferences for the disabled (Greece and Denmark this year). Their sound comes from instruments especially adapted for them – hand chimes, goatskin drums and aluminium bars, all tuned to different notes so that it is possible to play chords easily.

Their songs and shows are about the same themes that songwriters always choose. But their sheltered lives filter experiences differently. The show they devised themselves and toured earlier this year – The Dungeon Of Love – was, lead singer Pino Fruminto says, "about a group of people who've been loved too much". It was an allegory of the performers' own lives until now.

Hazy medication and humdrum day centres are still part of their lives, but the world of dressing rooms, recording studios and television cameras increasingly defines reality. They are the latest cross-over group, moving into the world of competitive entertainment. They emphasise that their new single is definitely not a charity record.

"It's made by a group of disabled people who've got together to do something for themselves," says Mark Williams, a musician who acts as co-ordinator for Heart 'n' Soul. "It's ironic that we've had the Knebworth concert with famous singers giving their services to benefit an organisation that does music therapy for the disabled, and here we have a group who are doing it for themselves. If there is another Knebworth, they should be invited.

The single, One in Four Theme, is distributed by the Total Record Company.   John Cunningham © THE GUARDIAN

Science, Technology and Society © D. Andrews, 1992

5 THE REPRODUCTION REVOLUTION

UNIT 5.7 Mass screening for inherited disorders

Inherited disorders cause considerable distress to the sufferers and their parents. In most cases the illnesses they cause are incurable and some, such as cystic fibrosis (see Unit 5.4), are fatal at an early age.

However there is some hope for the future, since scientists have developed the technology that could prevent all of them. There are three main approaches to achieving this.

1 Test the entire population to identify people who are healthy themselves but are unknowingly carrying a harmful gene that could affect their children.

2 Test the fetus before it is born (**prenatal testing**) and offer the mother an abortion if a severe inherited disorder is detected (see Unit 5.8).

3 Use IVF (*in vitro* **fertilisation**) techniques to obtain several embryos for couples who are known to be at risk of passing on an inherited disorder. Then test the embryos to select only healthy ones for implantation into the mother's uterus. This is discussed further in Unit 6.1.

In this unit you will investigate the first of these approaches. You will also explore the complex political, ethical and social issues that make the implementation of a mass screening programme a highly controversial problem.

SCREENING THE POPULATION FOR SICKLE CELL

Worldwide there have only been a few examples of mass screening programmes of the entire population to discover the carriers of a specific inherited disorder.

In the 1960s, some states in the USA introduced legislation that enforced the compulsory testing of all children who belonged to ethnic groups with a high risk of sickle cell anaemia – those of African and Asian origin (see Unit 5.3).

The aim of this programme was to discover which healthy individuals were carriers of the sickle cell gene. If they were told before they had children they could be alerted to the possible risks. This type of compulsory mass screening approach to the detection of potentially harmful recessive disorders was very controversial. Employers with racist motives began to use the results of the test as an excuse to refuse to employ black people who carried the gene. The outcry that followed led to a repeal of the laws.

Similar tests to detect carriers are available for a number of other harmful genes such as thalassaemia (which particularly affects Greeks, Italians and Cypriots) and Duchenne muscular dystrophy (DMD, which only affects males, though females can be carriers). Most recently the discovery in August 1989 of the exact location of the cystic fibrosis gene allowed a test to be developed within the space of a few months. As yet there has been no enthusiasm for introducing a compulsory testing programme to identify carriers of sickle cell or cystic fibrosis in Britain.

ACTIVITY

Your views on mass screening

1 One of the arguments used to support compulsory mass screening for carriers of harmful genes is that it would lead to a reduction in the number of children born with the inherited disease.
How would this reduction come about?

2 Discuss whether or not it is right that governments should have the power to introduce compulsory testing schemes like this.
Would it be more acceptable if the testing scheme were voluntary?

3 What would be a drawback of making testing voluntary rather than compulsory?

4 The American sickle cell screening programme involved removing children of high risk groups from their school lessons to provide blood samples for testing.
Discuss the likely psychological impact of this method of selection on the children.

5 Is it fair to expect a ten-year-old child to cope with the knowledge that he or she is a carrier of a harmful gene? Explain your views.

6 Suppose an engaged couple discover that they are both carriers of a harmful gene.
   a) How would they feel when they first discover this?
   b) What are all the possible options they could consider for their futures?
   c) Imagine you were in this situation. What option would you choose? Explain the reasons for your choice.

7 Explore this dilemma further by staging a role play between a couple who have just discovered that they are both carriers of the sickle cell gene. Imagine that they have been married for a year and that they were originally intending to have a family.
After the role play discuss the views and feelings displayed by the two people.

8 Consider this statement:
'Screening for harmful genes is the first step on the slippery slope towards total government control over those people who are allowed to reproduce.'
   a) How could compulsory mass screening programmes be used by a government in this way?
   b) What restrictions would be needed to prevent governments from going down this 'slippery slope'?

SCIENCE TECHNOLOGY AND SOCIETY

5 THE REPRODUCTION REVOLUTION

UNIT 5.8 Screening embryos and fetuses

During pregnancy the mother and her fetus are checked regularly by medical staff to ensure that they are both healthy. Most of the standard tests such as the mother's blood pressure or the position of the fetus in the uterus are straightforward. However more recent technological advances have produced a range of tests that are more controversial. They include amniocentesis to check for chromosome abnormalities (see Unit 5.5) and blood tests that detect the presence of high levels of the substance **alpha feto protein** (AFP) in the mother's blood, which could indicate spina bifida.

The controversy that surrounds these tests is partly due to objections to the use of abortion as a remedy when an abnormal fetus is detected. A new technique of embryo screening will avoid this particular ethical problem but raises others. One cell can be removed from a 16-cell embryo that has been formed with the aid of IVF (*in vitro* fertilisation) techniques, and tested for genetic abnormalities. This raises the possibility of screening out a wide variety of inherited features that might be considered undesirable. Some of the thorny issues that surround these two types of prenatal testing will be examined in this unit.

SCREENING FOR SPINA BIFIDA

One in 400 pregnancies produces a child with the congenital disorder spina bifida. ('Congenital' means that it is not inherited, but is a malformation that occurs during the development of the fetus in the uterus.) The vertebrae fail to enclose completely the delicate nerves in the spinal cord, which then become damaged during the pregnancy and birth. Such a baby will need extensive surgery to close this gap and the severity of the child's disability is difficult to predict. Although the child's intelligence is not impaired, he or she is likely to be wheelchair-bound and possibly incontinent.

Unlike amniocentesis, which is normally restricted to women over 35, a blood test for levels of the substance alpha feto protein is available for all pregnant mothers. It is known that if the fetus suffers from spina bifida it will produce large amounts of AFP, which can then be detected in the mother's blood. When high AFP levels are found, the fetus's spine can be examined using ultrasound imaging. (High AFP levels can also indicate healthy multiple pregnancies.) If spina bifida is diagnosed, the condition is explained to the pregnant woman, who is offered the choice of an abortion if she does not wish to have the child. There are many theories about the causes of spina bifida, but at present no specific cause has been identified.

SCREENING EMBRYOS

It is now possible to screen three-day-old embryos comprising only 16 cells to check for a number of genetic disorders. At this stage any single cell can be removed for testing without affecting the future development of the fetus. This allows affected embryos to be rejected and a healthy embryo to be selected in cases where the parents are at risk of passing on a harmful gene. This treatment was first used in 1990 to allow a woman who was a carrier of Duchenne muscular dystrophy to give birth to a healthy son.

ACTIVITY

Your views on the testing controversy

1. In a survey 25 per cent of doctors felt it was unnecessary to obtain the woman's consent for AFP screening.
   - Why do you think that the doctors took this view?
   - Do you agree with it? Explain your reasons.

2. What are your views on the following objections to prenatal testing for disorders?
   - All fetuses have a right to life regardless of their physical characteristics.
   - An emphasis on prenatal testing diverts attention away from the financing and development of treatments that would benefit babies born with handicaps.
   - Prenatal testing could be abused by parents who wanted to abort a child for trivial reasons, for example, if it is the 'wrong' sex.
   - In future our definition of an abnormal fetus could be widened to include minor imperfections.

3. In some cases doctors have argued that if a baby is born with severe physical handicaps such as spina bifida, the parents should be able to decide whether life-saving surgery should be given. If the parents decide against surgery the baby is given 'nursing care only'. This means that it is given the necessary treatment to relieve pain, but it will inevitably die.
   - Why do you think that this suggestion has been put forward by some doctors?
   - What are your views on this proposal?
   - Role play the discussion between a doctor and the parents who are considering this possibility for their newborn baby.

ACTIVITY

Your views on screening embryos

Suppose this technique was used to detect genes that do not cause a disorder with 100 per cent certainty, but simply increase the risk of suffering from a disorder. Such genes might include:
- genes that increase the risk of an acute mental illness such as schizophrenia
- genes that increase the risk of a heart disease that would require expensive corrective surgery
- genes that increase the risk of a severe form of diabetes that would require life-long insulin injections.

1. Why might a future government see an economic justification for insisting that all embryos should be screened for these three types of genes?

2. What would be the likely objections raised against such a proposal?

3. In the future, parents may be able to select embryos that have a desired set of their own characteristics. Do you see any objections to such a development? Justify your view.

5 THE REPRODUCTION REVOLUTION

UNIT 5.9 Genetic counselling

Genetic counsellors have an important role in helping people who may be carrying harmful genes. They have an extensive knowledge of human genetics, which they can use to calculate the risk of passing on an inherited disorder. It is their job to discuss these risks with clients and to present the options for future action. In this unit you will explore a case study that illustrates some of the techniques used by genetic counsellors and the problems they often encounter in their jobs.

HUNTINGTON'S DISEASE: A CASE HISTORY

Alice is 22 years old and has just completed her nursing training. When she was 15 her father died, after an illness that had steadily got worse over a period of ten years. It began with him slurring his speech and becoming more clumsy. This led to a steady deterioration of his mental abilities and physical coordination. In Alice's own words, 'he had become a living human vegetable'.

A genetic counsellor, Nasira Khan, interviewed Alice and found that several of her relatives had died from a similar disease. In the box is a record of their conversation. From this information Nasira was able to draw up a family tree or pedigree.

Investigating a family history

The following is a record of the conversation between Alice and the genetic counsellor, Nasira Khan.

Nasira: Do you know whether any of your relatives suffered from a similar illness to the one your father died of?

Alice: Yes! 12 months ago my Auntie Nancy died. It was terrible... just like my father – exactly the same symptoms, and she was only 43.

Nasira: Whose side of the family was this aunt on?

Alice: She was my father's younger sister.

Nasira: What about your mother's side?

Alice: She has three brothers but they are all healthy.

Nasira: Are your grandparents still alive?

Alice: Yes, apart from my father's mother. She died quite young... before I was born.

Nasira: Do you know what she died from?

Alice: Nobody seems to know. It's all a bit of a mystery.

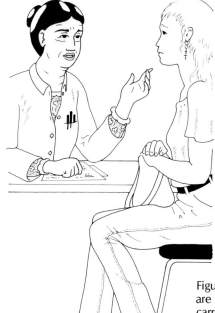

Figure 5.9.1   Nasira Khan's skills as a genetic counsellor are important in dealing with someone who may be the carrier of a fatal genetic disease.

ACTIVITY

Personal reactions

Read Alice's case history and then discuss these questions in small groups.

1. Use the information from Nasira and Alice's conversation to draw a family tree like the one shown in Figure 5.3.3.

2. Using this information, Nasira concluded that Alice was at risk from a fatal genetic disorder, caused by a rare single dominant gene, called Huntington's disease.
   How did Nasira know that it was a dominant gene that caused the disorder and not a recessive gene?

3. Nasira might draw a diagram similar to the one in Figure 5.3.2 to show Alice that she had a 50 per cent chance of suffering from Huntington's disease.
   Draw a diagram that she could use to show this.

4. Alice can have a simple test to discover for certain whether or not she has the fatal gene.
   How do you think Alice feels about having the test?

5. Nancy Wexler, a scientist working in Columbia University, New York, has become a world authority on Huntington's disease. The disease runs in her own family and she has a 50 per cent chance of carrying the fatal gene. But she has decided not to take the test. She says, 'When you think of yourself as a carrier of a trait that has no treatment, psychologically it has an impact.'
   If you were in this position would you want to have the test? Explain your answer.

6. If a person decides that they want to be tested it is standard medical practice to assess carefully the person's personality and present state of mind.
   What do you think this assessment might discover that would lead to a doctor refusing someone the test?

7. 'Doctors should not have the right to decide whether someone should be able to find out if they have the Huntington's disease gene.'
   Discuss this assertion.

8. Suppose Alice is told that she definitely has the Huntington's disease gene.
   How do you think her approach to life will change?

9. If someone has a history of Huntington's disease in the family, what is the earliest age that they should be offered a test to discover whether they have a harmful gene? Give your reasons for the age you choose.

Science, Technology and Society © D. Andrews, 1992

6 IMPROVING THE SPECIES?

UNIT 6.1 Test tube babies

In 1978 the birth of the world's first test tube baby, Louise Brown, gave hope to many infertile couples. But it also generated a great deal of controversy. This unit explains this technique and discusses its social and ethical implications.

IN VITRO FERTILISATION (IVF) AND EMBRYO TRANSFER

One in ten couples experiences infertility. The two most common causes are blocked fallopian tubes, which prevent sperm from reaching the egg, or a low sperm count, which drastically reduces the chances of any sperm surviving the long journey from the vagina to the ovary. The technique of *in vitro* **fertilisation** (IVF) is one way of overcoming these problems. After fertilisation the embryo is put into the uterus – **embryo transfer** (ET).

The woman undergoing the IVF treatment is first given hormones that cause a large number of eggs to ripen simultaneously in her ovaries. The doctor uses a laproscope to observe the ripe ova, which are each contained within a bubble of fluid (the follicle). The laproscope is a long, slim magnifying instrument, which is passed through a small incision in the abdominal wall. Several ova are then collected by gentle suctioning through a fine, hollow hypodermic needle and transferred to a petri dish, with a nutrient solution and a sample of semen. The fertilised eggs are allowed to develop for up to three days by which time they have reached the eight- or sixteen-cell stage. Two or three embryos are then placed in the woman's uterus through a tube passed through the cervix. This improves the chances of a successful implantation of at least one embryo.

> ### ACTIVITY
>
> #### Opposition to IVF
>
> Study the report reproduced from the *Guardian* newspaper of May 22 1984 before answering these questions.
>
> 1 What are Sir John Peel's major objections to test tube baby technology? Do you agree with him?
>
> 2 What do you suppose Sir John Peel was referring to when he accused the test tube baby specialists of ignoring the well-being of the children they had produced in these 'artificial ways'?
>
> 3 What two objections does Professor Duncan raise? How might these objections be answered by a supporter of the test tube baby technology?
>
> 4 Discuss the following statements about the treatment of infertility with IVF techniques.
>
> - The development of IVF technology is totally unnecessary when the world is desperately in need of solutions to the population explosion.
>
> - It would be better to encourage more positive attitudes to childlessness so that the demand for IVF treatment is reduced.

'Horrific' potential of test tube babies condemned

By Andrew Veitch,

Medical Correspondent

WORK on test tube babies is horrendous, and the potential of the technique is horrific, Sir John Peel, the Queen's former gynaecologist, said yesterday.

He stopped short of calling for a ban on the study of human embryos – he suggested a 10-day limit on growing embryos in laboratories – and agreed that there were some benefits, albeit to a "small number of women."

He was concerned that other species might be used to grow human embryos – scientists have already succeeded in using a horse to give birth to a zebra, and have created a sheep-goat by combining cells from the embryos of both animals.

As for those practising in vitro fertilisation, he said: "It is being done with a total disregard for the well-being and benefit of the child that is going to be produced. That is a terrible indictment of what professional doctors and scientists are doing. They are disregarding the effects of producing children in these artificial ways in order to satisfy the wishes of individuals."

Professor Ian Duncan from Glasgow University, who contributed to the report, was even more outspoken.

The most obvious abuse of the technique was the interference with sex discrimination (work is in progress on developing a sex probe for embryos), and using donated eggs and surrogate mothers could lead to breeding according to specification.

© THE GUARDIAN

6 IMPROVING THE SPECIES?

UNIT 6.2 Embryo research choices: Germany and the UK

Should IVF procedures be used to produce a number of embryos that are not implanted in the woman's uterus (see Unit 6.1)? If so, what should be done with these embryos? Is it morally acceptable to use them in scientific research? These controversial questions are considered in this unit, which gives you the opportunity to form your own opinions on these issues.

EMBRYO RESEARCH

Many scientists believe that the use of human embryos as material for research is essential for progress in solving a range of crucial scientific and medical problems. Such problems include investigations into the causes of infertility and miscarriage so that improved medical treatments for these conditions can be developed. It also covers research into inherited disorders, such as cystic fibrosis, and into abnormalities that occur during embryonic development, such as spina bifida (see Unit 5.8).

Yet the idea of experiments on human embryos disturbs many people who believe such research is both unnecessary and unethical. In June 1990 scientists in the UK were faced with the possibility that the research that they had been carrying out on human embryos would be banned by an Act of Parliament. In the German Parliament, known as the Bundestag, a similar law was debated in October 1990. However, the outcome of the political argument in the two countries was very different.

Box 1 gives *The Guardian* newspaper's report of the debate on embryo research in the House of Commons on 24 April 1990. An account of the Bundestag debate, which first appeared in the scientific journal, *Nature*, on 1 November 1990, is reproduced in Box 2.

ACTIVITY

Embryo research – the political debate

1. a) What is meant by a Government Bill?
   b) In what way is it different from an Act of Parliament?

2. MPs were given a free vote on the Bill, an opportunity that is granted very rarely.
   a) What do you think a free vote means?
   b) How does it differ from the normal House of Commons voting procedure?
   c) Why do you think the government took the unusual step of allowing a free vote on this Bill?

3. Sir Bernard Braine said in the debate: 'We are not talking about a cluster of cells. We are talking about a human life and it should be treated as such.'
   - Do you believe that the evidence supports his view?

4. Kenneth Clarke, then Minister of Health, said in the debate that he could not 'accept the early embryo as a human personality.'
   - What evidence did he present to support his view?
   - Do you agree with him?

5. Jo Richardson, the Labour spokeswoman on women's affairs, argued that if MPs had banned embryo research in the previous year a recent medical breakthrough would have been prevented.
   - What was she referring to?
   - Do you think her argument a good one in support of embryo research?

6. a) How does the German position on embryo research and IVF differ from the UK's?
   b) What historical differences between the two countries might underlie the distinct attitudes that have developed?

7. Why does the author of the *Nature* article believe that the German ban on experimentation is 'potentially dangerous'? Do you agree?

8. Some opponents of the German legislation believe that it does not go far enough. Why do they take this view and what additional restrictions would these people like to see?

Science, Technology and Society © D. Andrews, 1992

Proposal to legalise research during 14-day period after fertilisation wins 171 majority on free vote after emotional debate

MPs give overwhelming backing to medical research on embryos

**Martin Linton
and Nikki Knewstub**

THE House of Commons voted by a majority of nearly two to one last night to allow embryo research to continue up to 14 days after fertilisation under the Government's Human Fertilisation and Embryology Bill.

On a free vote, 364 MPs voted for the clause and 193 against, a majority in favour of research of 171. The Government took an officially neutral stance, although both the Health Secretary, Kenneth Clarke, and his deputy, Virginia Bottomley, said they would vote in favour.

Mr Clarke said the Government took no collective stance on the rights and wrongs of embryo research, but he personally would support the proposal to allow research up to 14 days after fertilisation.

The critical question for MPs before the free vote on the proposal in the Human Fertilisation and Embryology Bill was at what stage human life began, he said.

He did not believe a human being came into existence at the moment the egg was penetrated by the sperm; it was not possible to say at that stage which embryos would become a foetus and which the placenta.

Status as an individual could begin only at the stage where cells could be differentiated; this coincided with the appearance of the so-called primitive streak at 14 days, he said.

"For my own part, I can't accept the early embryo as a human personality with which I can identify as a person to whom the criminal law must give protection.

"That is in my case a layman's judgment, but I am influenced by the tiny size of the embryo and the undoubted fact that a high proportion of embryos at this stage perish naturally in any event."

David Alton (Lib. Dem. Liverpool Mossley Hill) intervened to ask whether Mr Clarke therefore accepted that at 14 days a developing human being was entitled to the full support of the law.

He replied that he thought the evolution of a human being was a steady and continuous process. "It's a very difficult matter to say at what stage do you have a citizen, a human being. At various stages fresh rights are acquired."

MPs had also been given impressive medical evidence that embryo research could help in the prevention or cure of infertility, congenital disease, miscarriages, misconception and abnormalities.

One couple in eight was infertile; further research could improve the success rate, so far disappointingly low.

Another tragic fact, said Mr Clarke, was that one in five pregnancies ended in miscarriage – about 100,000 a year – and much more work and research into the causes was needed.

Research could also lead to improved methods of contraception.

Jo Richardson, Labour's spokeswoman on women's affairs, said that in her view the emergence of the primitive streak at 14 days marked the beginning of individual development. Before that, it was a pre-embryo, a cluster of cells which might or might not develop into an embryo.

She rejected the "nasty and divisive" suggestion that those who supported research were lacking in respect and care for those with disabilities.

"Sometimes people point to the wonderful example handicapped people are to the rest of us; indeed they are, but true though that is they didn't ask to be born with a handicap.

"Life has dealt them a very cruel blow which they and their carers try to cope with. What is wanted is a choice, and pre-embryo research offers the prospect of that choice to particular groups of parents," she said.

If the Commons had banned pre-embryo research last year, MPs would have prevented the recent breakthrough at Hammersmith hospital which had allowed three mothers who were carriers of severe genetic disorders to have babies free from risk of handicap.

Sir Bernard Braine (C. Castle Point), the Father of the House, said: "We are not talking about a cluster of cells, we are talking of human life, and it should be treated as such." He said pro-Life MPs were not against all scientific research.

"What we do oppose is destructive or non-therapeutic research on human embryos of a kind which does not help the individual embryo to live, and to grow. We object to fatal experimentation. Those which kill the patient.

"And we hold that this is the only responsible and moral position to take."

He said in vitro fertilisation could continue even if there were a ban on research. "The question before us is not whether or not we should allow destructive or non-therapeutic research to continue. IVF will continue, and will not be stopped." Detecting defective embryos would still be allowed.

Sir Bernard said: "The buck stops here. We have to make up our minds about it." If much could be done without using human embryos in research, why were so many so eager to conduct such research?

He said hundreds of thousands of eggs would be needed, and women would be treated with super-ovulatory drugs. They would be treated as guinea pigs. He referred to Nazi experimentation, where the end justified the means, and pointed out that the West German parliament had recently voted to ban experimentation.

Peter Thurnham (C. Bolton NE) said that what was being discussed was a cluster of undifferentiated cells, which yet had the primitive streak, the start of individuality.

Women using contraceptives such as coils lost millions of fertilised embryos. It was also not possible at the moment to keep an embryo alive beyond nine days outside the womb.

It was an immoral argument to favour IVF but not experimentation. Over 300 experiments were carried out before the birth of Louise Brown, the first IVF baby.

A bizarre suggestion came from **Sir Trevor Skeet** (C. Bedfordshire N) that cost might be a factor in support for human embryo research.

"The expense of having monkeys and maintaining their upkeep is becoming prohibitive. In contrast the human embryo is readily accessible and relatively inexpensive."

Edwina Currie (C. Derbyshire S) said she had been through a miscarriage and it was a dreadful experience.

Research needed strict controls, but scientists could not turn their backs on the knowledge they had gained through research. "Would it not be a good thing if the haemophilia gene vanished, if muscular dystrophy became a thing of the past?" she said.

Rosie Barnes (SDP Greenwich) said that in the 13th week of pregnancy she had German measles. After seven weeks of tests she was given 24 hours to decide whether to have an abortion in the 20th week.

She decided to have the child and now has a 17-year-old son who has only slightly impaired hearing. "But that seven weeks was a very miserable time. That decision whether to go ahead with the pregnancy knowing that you might have a severely disabled child, or that you might be aborting a child that was quite well and might have a reasonable hope of happiness and fulfilment, was a dreadful decision and I wouldn't wish it on anyone.

"For anyone to be able to make that decision in advance of the baby growing in their womb must be a great step forward."

Mrs Bottomley said the Government was confident the new authority would be able to impose strict control on research and would have powers to revoke, vary or suspend licences. These would be allowed only for specific research into infertility, miscarriage, abnormalities or contraception.

© THE GUARDIAN

EMBRYO RESEARCH

Germany turns clock back

Munich

THE German Bundestag last week passed a law making research on human embryos a criminal offence punishable by up to five years in prison and imposed severe restrictions on the use of *in vitro* fertilization (IVF). The law, which also includes limits on changes in the human germ line and other invasive genetic procedures, gives Germany the world's strictest embryo research regulations. The new law takes effect on 1 January 1991.

The ban sets a potentially dangerous precedent for research in Germany by calling into question the constitutional protection of freedom of research. But efforts by research organizations to incorporate exceptions to the ban were rebuffed by the shared view of all political parties that such research is unethical.

The legislation imposing the ban specifies that embryos may not be created for any purpose other than implantation in a woman's uterus. The number of egg cells that may be fertilized for any one IVF operation is limited to three. The use of embryos for any other purpose including research is to be considered a criminal act.

Significantly, although the German law forbids research involving the removal of so-called 'totipotent' cells (which could be used to create a new embryo if removed at the 8-cell stage of the dividing embryo or earlier), it will allow prenatal diagnostic testing of later stages. Physicians can test for genetic diseases on cells removed from the 32-cell or 64-cell stage, known as the morula, without harming the embryo. If a disease is detected, the embryo can be discarded before implantation.

According to Henning Beier, a professor of anatomy and reproductive medicine at the Technical University in Aachen, the law does not present an "insurmountable barrier" to embryology and diagnostics research. Beier says that almost all embryology research in Germany is carried out using animals.

Two large German research organizations, the Max Planck Gesellschaft and the Deutsche Forschungsgemeinschaft, declared in 1988 that they would observe a self-imposed moratorium on embryo research until there was a law.

Both research organizations supported sections of the law that ban the cloning of human beings and the creation of human-animal chimaeras. But some researchers wanted to allow exceptions to a ban on manipulating the germ line in cases where gene therapy might be of benefit. Harald zur Hausen, director of the German Cancer Research Centre in Heidelberg and an outspoken opponent of the law, said that pressure would soon grow to revise the law in order to allow gene therapy.

The new law would have banned completely the use of preimplantation diagnostic tests to select the sex of the embryo to be implanted. But at the last minute the government chose to add an exception allowing selection against embryos carrying severe sex-linked disorders such as Duchenne muscular dystrophy in which case the son of a mother who is a carrier has a 50 per cent chance of developing the crippling disease, whereas daughters are rarely affected. Therefore male embryos are not usually implanted.

The opposition Social Democrats and Green Party voted against the law because it did not go far enough. The law gives "legislative blessing to eugenic measures" in Germany [for the first time since 1945], says Anna Waldschmidt, a spokeswoman for the Green Party.

Green parliamentarian Marie-Luise Schmidt strongly criticized the government during a Bundestag debate on 24 October for allowing embryos carrying genetic diseases to be discarded.

Schmidt said that this provision comes dangerously close to the selective measures included in the Nazi laws for the prevention of genetically diseased offspring. "People who were born with muscular dystrophy are apparently meant to find themselves in this law as the objects of a . . . negative selection that defines them as worthy of extermination", she said.

The Green Party opposes IVF in general as a "massive experiment on human subjects," because the techniques do not allow for a high rate of success. Instead, the party urges support for more research into the causes of infertility and into solutions other than IVF.

By Steven Dickman. Reprinted by permission from NATURE vol. 348 p. 8 Copyright © 1990 Macmillan Magazines Ltd.

6 IMPROVING THE SPECIES?

UNIT 6.3 Freezing embryos

The technology of embryo freezing (**cryopreservation**) has already had a major impact in animal husbandry, by allowing farmers to select stock with a particular set of characteristics. Animal conservationists have also seen its potential and suggested that embryo banks of endangered species should be maintained in case the wild population falls to dangerously low levels. These applications of this technology are not controversial, but when human embryos are involved a host of difficult issues arise and these are explored in this unit.

HUMAN EMBRYO BANKS

The four-day-old human embryo, which consists of a bundle of 64 cells about 0.1 mm across, can be stored at a temperature of $-196\,°C$ in liquid nitrogen for several years. This sub-zero temperature arrests its development until it is thawed out, when it can be implanted in the original mother or in another woman.

The world's largest 'embryo bank', at Monash University in Melbourne, Australia, takes surplus embryos from the many IVF procedures that are carried out at the hospital there. The bank was originally set up because the university's ethics committee decided that freezing embryos was more acceptable than disposing of them. The bank can store a number of embryos produced by a couple from a single IVF treatment, which can then be implanted into the mother's uterus when future pregnancies are required.

Total fertility control – the ultimate in family planning

Planning when your children are born has always been a crucial decision. Until now you have had to live with a host of uncertainties.

- How will my career be affected if I take time out to have a baby?
- Will the contraceptive method I'm using be 100% reliable?
- Will I be able to get pregnant as soon as I stop using contraception?
- What will be the risks of my child being born handicapped if I wait until I'm well established in my career in my late thirties?

But now you can achieve the ultimate in family planning through the unique services of Total Fertility Control. We give you the option of choosing exactly the age when you want to have children with no risk of unwanted pregnancies or a handicapped child. This is what our service offers:

- While you are still a young woman a number of eggs are removed from your ovaries.
- These eggs are fertilised in our laboratory using sperm donated by your partner.
- The resulting embryos are checked to ensure that they are free from inherited disorders; then frozen and stored for as long as you require.
- At a time in your life that suits you, your embryo is unfrozen and implanted in your uterus.
- Secure in the knowledge that you have already conceived all the healthy babies that you will desire, you can be sterilised to ensure that there is no possibility of your planned family being disrupted by a 'little mistake'.

ACTIVITY

Your views on frozen embryos

1. Several countries have put a limit on the length of time that frozen embryos can be stored.
   - Why is such a time limit imposed?
   - In the UK the time limit is five years. Do you believe this is a suitable time limit?
   - Do you think it should be longer or shorter? Explain your point of view.

2. What should be done if both the parents decide that they no longer wish to keep the frozen embryos?

3. What might be the objections to the storage of frozen embryos?

4. What are your views on the desirability of frozen embryo banks?

5. a) If a frozen embryo is donated to an infertile couple who are not its biological parents, what problems might arise as the child grows up?
   b) How do you think these problems could be resolved?

6. In June 1984, Mario and Elsa Rios, a wealthy Los Angeles couple, were killed in a plane crash. They left two frozen embryos.
   - Who do you think should now be responsible for these 'orphan embryos'?
   - Do the embryos have a right to be implanted and therefore a right to life? Justify your answer.
   - Should they be donated to another infertile couple? If this is done, do the embryos have a right to inherit their biological parents' fortune?

7. What legislation, if any, do you think should be introduced to control the operation of frozen embryo banks?

8. The box contains an advertisement for a fictitious company, 'Total Fertility Control'. The service described is entirely feasible, using the present technology.
   - Do you think there is demand for such a service? Explain your answer.
   - If a similar company were set up in the future, would there be pressure to ban its activities? If so, what would be the likely objections?
   - If such a company could offer this service, would you want to make use of it? Give your reasons.

SCIENCE TECHNOLOGY AND SOCIETY

6 IMPROVING THE SPECIES?

UNIT 6.4 Surrogate motherhood: rent a womb?

The use of a surrogate mother as a solution to a couple's infertility is not an invention of modern technology. Throughout history there are many recorded cases of a woman agreeing to be inseminated by another woman's husband and then handing over the child to the couple. However, since the early 1980s the use of IVF techniques has greatly expanded the potential for different surrogacy arrangements, which are detailed in Box 1. This unit explores the ethical and legal problems that are raised by this new technology.

Britain's first surrogate pregnancies were reported in March 1984 (see Box 2). They had been arranged by a commercial firm that made a profit from providing this service. A year later, in 1985, commercial surrogacy was outlawed in the UK by an Act of Parliament. However non-commercial surrogacy was still legal. If doctors brought together infertile couples and women prepared to act as surrogate mothers they were not breaking the law, provided that no financial transactions were involved.

After nine months of pregnancy and the birth the surrogate mother may have formed a close attachment to the child. Despite this, she is required to give it up to the couple who have commissioned her. This is clearly very difficult and there have been some cases where the surrogate mother has tried to retain custody of the child (see Box 3).

ACTIVITY

Your views on surrogacy

Read the three different surrogacy arrangements in Box 1, the article in Box 2 and the case of 'Baby M' described in Box 3 before discussing the following questions.

1. What are your views on commercial surrogacy? Was the UK Parliament right to ban it?

2. Do you consider the three examples of surrogacy arrangements described in Box 1 to be ethically acceptable?

3. In which of the three arrangements does the surrogate mother have the greatest claim on the child?

4. What disputes over legal parenthood might arise in each of the different surrogacy arrangements?

5. What legal restrictions, if any, should be imposed on surrogacy arrangements?

6. Were you surprised by the outcome of the Baby M case?

7. What changes would you recommend to avoid a repetition of the Baby M case?

8. Consider the case of a woman who is not infertile but wants to use a surrogate mother as described in case 1, Box 1.
   - Suggest possible reasons for a woman wanting to enter a surrogacy arrangement to avoid pregnancy and birth.
   - Discuss the desirability of allowing this type of surrogacy.
   - What are the wider implications of this development for the future of human reproduction?

1. Different surrogacy arrangements

The different arrangements for surrogate motherhood are as follows:

1 The infertile couple donate both egg and sperm, which are fertilised *in vitro*. Their embryo is then transferred to the surrogate mother's uterus.

2 Infertility is due to the woman's failure to ovulate. The surrogate mother is inseminated with sperm from the father.

3 Both the man and woman are infertile and they select a frozen embryo from an embryo bank, which is implanted in the surrogate mother's uterus.

2. First surrogate births are on the way

Britain's first surrogate pregnancies were announced last night by an American agency that pays women to carry babies from conception to childbirth on behalf of would-be mothers who suffer from infertility.

Two British women have undergone artificial insemination and accepted a £6,500 fee to carry the babies for nine months and surrender them to the couples who commissioned the pregnancies.

The couples involved have agreed to pay more than £13,000 to cover the fee to the surrogate mother, a similar sum to the agency and all medical fees.

The practice of surrogate motherhood has been condemned by the BMA because of the difficulties and uncertainties involved. It may be outlawed this summer after the Warnock Committee reports to the Government.

3. The case of 'Baby M'

In 1985, William Stern and his wife signed a contract with a commercial surrogacy agent in New York. They paid $10,000 to obtain the services of a surrogate mother, Mary Beth Whitehead, who was inseminated with William Stern's sperm. After the birth the child was handed over to the Sterns as agreed in the contract, but Mary Beth Whitehead then changed her mind and began a legal battle to gain custody.

After two years of intensive legal argument the Supreme Court of New Jersey made the final decision. Mary Beth Whitehead was ruled to be the legal mother of 'Baby M' and the original contract was declared void. Nevertheless, the custody of 'Baby M' was granted to the Sterns.

6 IMPROVING THE SPECIES?

UNIT 6.5 Genetic engineering

When the chemical structure of DNA was discovered, it opened up the possibility of taking a suitable organism and artificially modifying and manipulating its genes. The idea of genetic engineering was born. This unit explores some applications of genetic engineering and the wider social questions raised by the need to control this new technology.

GENE TRANSFER

The most important of the genetic engineering techniques developed allows scientists to remove a specific gene from one cell and insert it into a different cell – a process called **gene transfer** (see Figure 6.5.1). When the two cells come from individuals of different species, such as humans and bacteria, exciting new possibilities are opened up. Gene transfer has already been applied to a wide variety of problems in the pharmaceutical and food manufacturing industries. It also has a crucially important role to play in future treatments of inherited disorders (see Unit 6.6).

HUMAN GENE TRANSFERRED TO A BACTERIUM

People with one type of diabetes must inject themselves regularly with insulin to keep their blood sugar levels under control. Insulin is a hormone produced in the pancreas and traditionally animal insulin has been extracted from the pancreatic tissue of slaughtered cattle and pigs for diabetic use. Although animal insulin works well in controlling human diabetes, there are undesirable long-term side effects that doctors believed might be avoided if human insulin could be used instead.

The only way this could be achieved was by gene transfer. In 1978 the human gene that produces insulin was successfully transferred into a bacterium. Although it took some time to develop a cheap production process, by 1990, 80 per cent of diabetics were using insulin that had been produced in this way.

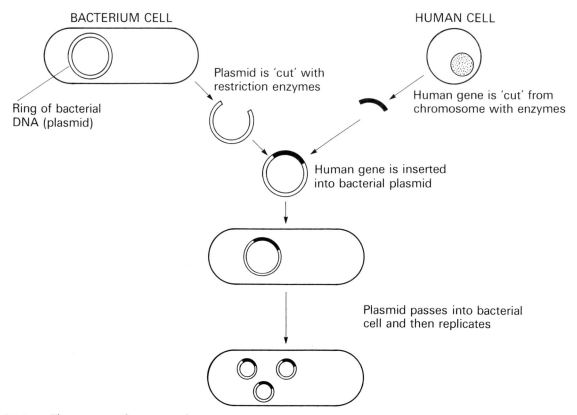

Figure 6.5.1   The process of gene transfer.

CONTROLLING THE GENETIC ENGINEERS

The techniques of genetic engineering give scientists a significant new responsibility. They now have the power to form **genetically modified organisms** (GMOs) with novel features that do not occur in the original species. This does not mean that they can design organisms from scratch in the same way as an architect designs a house. However the introduction of even a minor genetic change may carry an unknown risk.

In some cases it is intended that the GMO will be released into the environment and there are fears that its ecological impact could be unpredictable and potentially disastrous. There is a real distinction between the way a release of harmful chemical and the release of a genetically modified organism can affect ecosystems. A chemical pollutant, such as an oil spill, will seriously disrupt an ecosystem, but it can be treated and will eventually disperse. A genetically modified organism would not be so easily dealt with. Once it was released it would multiply rapidly and would be almost impossible to contain if it began to have harmful ecological effects.

ACTIVITY

Controlling the release of GMOs

Read the case studies of the development of two types of GMO in Boxes 1 and 2 before answering these questions.

1. Explain how GMOs could be an improvement on conventional chemical insecticides.

2. Even if an effective GMO insecticide can be developed, industry may still decide not to market it.
   - Suggest possible reasons for this.

3. In some countries there is a ban on the release of GMOs into the environment. Why do you think this is?

4. In other countries an official government committee has been set up to control the development and environmental release of GMOs.
   - Do you think members of the public should be represented on this committee?
   - What sort of powers should they be given?

1. Modified fungus to be deployed as an insecticide

Although farmers are aware that chemical insecticides pose a risk to wildlife, they have had to use them to avoid large-scale damage to their crops when no effective alternative has been available. Soon they may have the option of using a genetically modified fungus as an insecticide that has less risk of disrupting ecosystems.

Scientists searching for the ideal species of fungus had over a hundred to choose from. These species attack insects by reproducing rapidly inside their bodies and killing them within a few days. In selecting the most suitable fungus, scientists took the following three factors into account:

- It should be possible to mass produce the fungus in large quantities (tonnes).
- The fungus should store well over long periods before it is used.
- After storage the fungus should be easily activated when it is applied to crops.

Dr Paul Reinecke, Head of the Institute for Biotechnology of the Agrochemical Sector at Bayer, the company developing this product, explained how economic factors would also be taken into account before they decide to market it.

'This product has to be able to compete with chemical pesticides in terms of price, effectiveness and the level of application needed. Otherwise it won't stand a chance in the market.'

from *research*, 1989

2. NoGall, no thanks?

In March 1989 a unique pesticide, 'NoGall', was launched in Australia. NoGall was the world's first genetically modified bacterium to be used as a pesticide. The bacterium used was a strain of the species *Agrobacterium tumifaciens*. This harmless strain produces an antibiotic that kills the closely related pathogenic strain responsible for 'crown gall disease' – a type of plant cancer. Vast quantities of the world's fruit harvest, valued at $150 million, are destroyed by this disease every year.

To make the bacterium effective a specific section of its DNA had to be snipped out using genetic engineering techniques. Some conservation groups were alarmed that the release of an altered strain of bacteria might have unpredictable effects on ecosystems. However Australia's official regulatory body has approved its release and it is now on sale.

Science, Technology and Society © D. Andrews, 1992

SCIENCE TECHNOLOGY AND SOCIETY

6 IMPROVING THE SPECIES?

UNIT 6.6 Gene therapy

Since genetic disorders are the result of a faulty gene the obvious treatment is to replace this gene with one that functions correctly. But what are the risks of such treatments? Which genes are candidates for replacement or improvement? Could the treatments be open to abuse? You will consider these questions in this unit.

During the 1990s there will be considerable effort put into devising gene therapy for the common genetic disorders such as cystic fibrosis. Although such a breakthrough will be welcomed, there is concern that these techniques could lead to a demand for unacceptable types of gene transfer that may have to be regulated by law.

THE FIRST GENE THERAPY TRIALS

In September 1990 the first trials of gene therapy on humans began in the USA. The scientists who wished to carry out these experiments had had to wait three years for the go-ahead from a specially appointed panel of experts. The panel eventually decided that the risks of the trials were low enough to be acceptable.

The first ever gene therapy patient was a four-year-old girl who was suffering from adenosine deaminase (ADA) deficiency. This is a rare and fatal inherited disorder in which the white blood cells are unable to make the enzyme ADA, which is vital for fighting infections. A sample of white blood cells (T-lymphocytes) was removed from her bloodstream. Copies of the normal gene, capable of producing ADA, were obtained from the T-lymphocytes of a healthy donor and inserted into the patient's T-lymphocytes, using the technique of gene transfer (see Unit 6.5). The T-lymphocytes were then injected back into the child, where they began to produce ADA.

THE SAFETY OF GENE THERAPY

The insertion of new genes into human cells involves a degree of uncertainty. There is the possibility that the new gene will not act in the way it is expected to. For example, it may affect the ways that other genes are expressed and possibly lead to cancer. The scientists carrying out human gene transfer trials argue that the techniques are only being tried on patients whose illnesses cannot be treated successfully by other means. They also emphasise that the genetic changes that they make only affect a specific population of body cells. This is why it is described as **somatic cell therapy**. The genetic alterations have no effect on the cells that will form the gametes (the germ line cells). This means that the inserted gene cannot be passed down to future generations.

ACTIVITY

Gene therapy out of control?

Imagine that the year is 2020. A wide range of genes are now available for transfer into patients at their request.

The public is concerned that the gene therapy business is going too far. A campaign to bring a halt to gene therapy has been launched.

Consider the following somatic cell gene therapy treatments that might have been developed by the year 2020. Would you like to see any of them banned? Give your reasons.

1. A gene that enters the lining of blood vessels and releases a chemical that reduces the levels of fat and cholesterol in the blood.
2. A gene that produces a hormone with a tranquillising effect on the emotions.
3. A gene that produces a hormone that has a stimulant effect, improving the person's concentration.
4. A gene that produces a substance that provides protection against lung cancer.
5. A gene that has the effect of increasing an athlete's stamina.
6. A gene that if introduced in childhood has the effect of increasing the person's adult height.
7. A gene that releases a pigmented substance that permanently changes the person's skin or hair colour.
8. A gene that provides protection against the effects of radiation.

UNIT 6.7 Who owns the human genome?

The 100 000 genes found in our cells make up the human **genome**. At present the exact structures of only a few genes have been discovered, but when the structures are known, exciting new avenues of research open up. For example, when the cystic fibrosis gene was discovered, great advances were made in the search for a cure (see Unit 5.4). Clearly this knowledge is very valuable, but can the scientists who discovered it be said to 'own' it? This unit explores this issue by highlighting the efforts of one scientist to obtain copyright control over the human genome.

THE HUMAN GENOME PROJECT

In 1988 a group of American scientists, led by the eminent co-discoverer of DNA, James Watson, announced that they would spearhead a worldwide research project to make a complete analysis of the human genome. The Human Genome Project, as it was called, would take an estimated 30 000 person years – first to find the position of every gene using mapping techniques and then to analyse the structure of each gene in terms of a sequence of bases (see Unit 5.2). Great hopes are pinned on this project. In particular the mapping and sequencing of many harmful genes could provide a tremendous boost to the development of new tests and treatments for genetic disorders.

However the Human Genome Project has its critics. Some see it as a waste of money. They also fear it will lead to an expansion of genetic screening to detect healthy carriers of harmful genes. This could create 'genetic ghettos' of people who function perfectly well but who carry the stigma of being labelled as less than perfect.

THE GENOME CORPORATION

In 1989 Walter Gilbert, a professor of molecular biology at Harvard University in the USA, saw that the Genome Project offered a golden business opportunity. He announced that he was intending to set up his own private company, the Genome Corporation, to begin sequencing the human genome. To start up the company he was seeking $10 000 000 venture capital from investors.

To provide ongoing finance for his ambitious project, Gilbert planned to copyright each section of DNA as it was sequenced in the Genome Corporation's laboratories. Scientists from other universities or companies who wished to use a particular section – for example, in the development of treatments for a genetic disorder – would have to pay a fee.

In response to critics of his plan, Gilbert is quoted as saying:

> 'A lot of people have made a fuss about this but I don't see any problem. It's like writing a book. People can read the information but they will have to pay if they want to use it elsewhere.'

ACTIVITY

Copyrighting the human genome

1 How many scientists would have to work full time on the project if they were aiming to complete it in 20 years?

2 Suggest several possibilities for the source of finance for this mammoth research task.

3 What protection does the author of a book gain from the copyright laws?

4 In what ways is the author of a book
   a) similar   b) dissimilar
   to a scientist who has sequenced a gene?

5 In your view, does this comparison justify Gilbert's position on copyrighting human genes? Explain your reasons.

6 One researcher has said:

> 'What happens when Gilbert's corporation stumbles across a really crucial discovery like one that could lead to a cure for cancer? Will they sit on it and later try to make millions when it could save thousands of lives if widely released? It's very worrying.'

- Do you believe this fear is justified? Explain your point of view.

SCIENCE TECHNOLOGY AND SOCIETY

7 FEED THE WORLD

UNIT 7.1 Good for you?

How do you know if you are eating a healthy diet? We all have an idea of foods that are 'good for us' and those that are 'bad for us', but how accurate are these judgements? It is known that most people in the UK eat an unhealthy amount of fat in their diet. Yet someone who excludes all fat from their diet will soon die. Very high quantities of certain vitamins and minerals can be harmful to health, yet in the correct quantities they are essential. It is best to aim for a balanced diet that has the right proportions of different foodstuffs. It is generally agreed by nutritional scientists that most people eat too much salt, fat and sugar and not enough fibre and we should modify our diets accordingly.

---

medium-sized banana

three fish fingers

half a honeydew melon

Mars® bar

small packet of crisps

can of fizzy drink (not sugar-free)

medium-sized apple

slice of bread and butter

jam doughnut

fried egg

three boiled carrots

digestive biscuit

small pot of strawberry yogurt

ACTIVITY

Achieving a balanced diet

1 There are six groups of food types that must be included in a balanced diet. Name each one and the reasons that they are needed.

2 It is important to achieve a balance between the energy value of the food in your diet (energy input) and the energy that you need for activity (energy output). What happens if:
   a) energy input exceeds energy output
   b) energy output exceeds energy input?

3 Study the list of foods in the box and then rank them according to the size of the energy (calorific) value that you think they have. Compare your rank order with a partner, and discuss the differences. Agree jointly on a rank order and check your answer (see Teacher's Notes).

4 Rank the foods again, from the one you most enjoy at the top to the one you least enjoy at the bottom.

5 Discuss the differences between the rank order you gave in answer to question 4 and the calorific rank order.

6 Use a suitable computer program to analyse the nutritional value of your own diet. What foods do you need to eat more/less of?

7 A food scientist once said that crisps were not unhealthy because they contained as much vitamin C as an apple. Do you think that this is a good argument for recommending crisps rather than apples as an essential part of a balanced diet?

8 Find out how your long-term health could be affected if your diet was unbalanced in each of the following ways. Suggest which foods could be added to or excluded from the diet to correct each type of imbalance.
   - too much fat
   - too little calcium
   - too little fibre
   - too little iron
   - too much sugar
   - too little protein
   - too much salt

Science, Technology and Society © D. Andrews, 1992

7 FEED THE WORLD

UNIT 7.2 Food additives

Chemicals have been added to food for most of human history. The addition of salt or vinegar was a valuable method for preserving food before the invention of fridges and freezers in the twentieth century. Nowadays the range of additives used by the food processing industry is enormous – an estimated 3500 different ones are currently being used and of these about 500 have been given an E number. They perform a wide variety of functions (see Table 7.2.1) but there has been some concern that certain additives may have undesirable health risks.

Table 7.2.1   Food additives in common use.

Additive	Function	Examples
Preservative	Kills bacteria and prolongs the time food can be stored before being eaten	Benzoic acid, nitrites
Colour	Improves attractiveness of food	Beta-carotene, tartrazine
Flavouring	Adds flavour to processed foods to improve their taste	Caramel
Flavour enhancer	Brings out the flavour of other ingredients in the food	Monosodium glutamate (MSG)
Emulsifier	Prevents the separation of a mixture of liquid ingredients	Stearyl-nitrate
Stabiliser	Works with emulsifiers to improve the texture of food	Citric acid
Sweetener	Adds sweetness to foods and drinks without using sugar	Saccharin, Aspartame
Antioxidant	Prevents the oxidation of fats, which would cause them to taste rancid	Butylated hydroxytoluene (BHT)
Nutrient	Improves nutritional value to offset the losses during processing of the food	Vitamin C, iron, calcium

ACTIVITY

Differing views on food additives

1 Survey the uses of additives in foods by examining the labels on food packages. For each of the additives listed in Table 7.2.1, find an example of a food that contains it.

2 Find out about the controversy surrounding the food additive tartrazine (E102), which gives a yellow colour to foods and drinks. This additive has been used a great deal less since it was found to cause skin rashes in people who are allergic to it. Find out what natural colouring is now used instead of tartrazine.

3 Some scientists believed that they had evidence that tartrazine was responsible for causing hyperactivity in children. This evidence was queried on the grounds that other factors could have been involved. How would you carry out an investigation to test whether a child's behaviour was affected by the level of tartrazine in the diet?

4 In pairs discuss the following statements about food additives.
- Additives are essential because they preserve food for longer, reducing the risk of bacterial growth and food poisoning.
- Food itself is only a concoction of chemicals. What is wrong with adding a few more in the form of useful additives?
- Preservatives such as nitrites should be banned because the health risk from these additives is greater than the risk of getting food poisoning.

5 An antioxidant, BHT, has been banned from processed babyfoods but not outlawed altogether. Suggest a possible explanation for this position.

7 FEED THE WORLD

UNIT 7.3 Food advertising and labelling

Like any other product, food is packaged and advertised to try to get people to take it from the supermarket shelves and add it to their shopping baskets. Labels give the food manufacturer the opportunity to make the product look attractive, so you will want to buy it. But food labels can also provide useful information for the shopper. By law, any food that is processed must bear a label that includes the following:

- the name and address of the company that packages the food
- an accurate and truthful description of the food contained in the packaging
- the mass or volume of food contained within the package
- a complete list of the food's ingredients, in order of mass, including all food additives.

In addition, although it is not a legal requirement, some food labels contain more detailed nutritional information (see Figure 7.3.1).

ACTIVITY

Match the label

1. Why could the name and address of the food's packager be useful to the consumer?

2. Study the lists of ingredients on the labels in Figure 7.3.2. Try to identify which label belongs to each of the following foodstuffs:

    packet soup

    cereal bar

    Branston pickle

NUTRITION INFORMATION		
TYPICAL VALUES		
	PER 100g (3½ oz)	PER SERVING 60g (2 oz)
ENERGY	385 k. CALORIES	230 k. CALORIES
	1630 k. JOULES	975 k. JOULES
PROTEIN	10.5g	6.3g
CARBOHYDRATE AVAILABLE	58.7g	35.2g
TOTAL FAT	13.8g	8.3g
of which POLYUNSATURATES	2.6g	1.6g
SATURATES	4.5g	2.7g
DIETARY FIBRE	5.7g	3.4g

Figure 7.3.1 Food labels can often give useful nutritional information.

INGREDIENTS WHEN RECONSTITUTED:- CARROTS, ONIONS, SWEET RED PEPPERS, STARCH, FRENCH BEANS, PEAS, HYDROLYSED VEGETABLE PROTEIN, SALT, CELERY, FLAVOUR ENHANCERS (MONOSODIUM GLUTAMATE, SODIUM 5'-RIBONUCLEOTIDE), LEEKS, HYDROGENATED VEGETABLE OIL, YEAST EXTRACT, SPICES, HERBS.

(a)

INGREDIENTS: VEGETABLES IN VARIABLE PROPORTION (CARROTS, CAULIFLOWER, GHERKINS, MARROWS, ONIONS, RUTABAGA, TOMATOES), SUGAR, VINEGAR, DATES, SALT, APPLE, MODIFIED STARCH, LEMON JUICE, CARAMEL COLOUR, SPICES, ACETIC ACID, GARLIC EXTRACT.

(b)

INGREDIENTS

Rolled Oats, Chocolate Chips (16.9%), Glucose Syrup, Crisped Rice, Vegetable Oil, Dextrose, Sugar Syrup, Almonds, Unrefined Sugar, Coconut, Malt Extract, Honey, Vanilla & Salt.

(c)

Figure 7.3.2 Labels from a cereal bar, a jar of pickle and a packet of soup.

ACTIVITY

TV food and drink advertising survey

With a partner, plan a survey of the different food and drink products advertised on commercial television channels over a period of one week.

Allocate certain times and channels between you so that you do not overlap, and agree on a system for recording the information about each advertisement.

Design your survey so that the data you collect can be analysed to answer the following questions.

1. What proportion of total advertising time do food and drink advertisements take up?
2. What proportion of advertising time does each type of foodstuff take up?
3. Suggest possible reasons for differences in the time allocations.
4. Give some examples of food brands that claim to be healthy, nutritious or natural. Do you think these claims are justified?
5. Which age groups are different food brands being aimed at?
6. Certain foods are rarely advertised on television. Give some examples and suggest why television is not regarded as a suitable medium for boosting their sales.
7. What images and atmospheres are associated with
   a) chocolate
   b) crisps and snacks
   c) beer?
8. What is the likely influence of certain celebrities being used to endorse particular foods or drinks, for example, the model Jerry Hall advertising Bovril®?
9. The Advertising Standards Authority (ASA) is responsible for investigating complaints from the public about advertisements that do not meet the requirements of truthfulness, fairness and accuracy. The *ASA Case Report* is published regularly and provides a summary of all the complaints they have investigated. Study a copy to find examples of food advertising that have been investigated by the ASA. Are there any examples you have come across that contravene the ASA guidelines?

ACTIVITY

Food and drink research project

Gather together information on a particular food or drink that interests you and present it to the class in the form of a talk or a wall display. Your research could include the following:

- Write to the company that produces it to find out the details of the food's history and its manufacturing process. Some foods such as fish, meat and milk have national marketing organisations that will also provide useful information.
- Describe the food's nutritional value and its importance as an ingredient for cooking attractive dishes.
- Discuss whether there is a possibility of a risk to health if the food or drink is consumed in large amounts.
- Survey a sample of people to find out how popular it is.

7 FEED THE WORLD

UNIT 7.4 Mycoprotein

The world is faced with a serious protein shortage. The climates of many developing countries are unsuitable for raising livestock for meat production and their poor soils cannot be used to grow protein-rich plants such as beans. One possible solution is to develop the potential of microbes that are rich in protein (Single Cell Protein, SCP). After twenty years of research, the technology to grow and process microbes to make food products is now well-developed. But will people buy food that is derived entirely from microbes? This unit looks at one such food – the fungal protein (mycoprotein) marketed in the UK as Quorn®.

EDIBLE FUNGI

Mushrooms are a type of fungus that have traditionally formed a part of our diet. But another type of fungus – the filamentous fungus – does not at first sight appear to be as promising as a potential food source. This group includes such species as pinmould (*Mucor*), which readily grows on moist bread. In the 1960s the multinational food company, Rank Hovis MacDougall (RHM), undeterred by the apparent difficulties, began to look for a suitable species. They eventually selected the filamentous fungus *Fusarium*, which grows well on a solution of simple sugar and ammonium ions, doubling in mass every five hours at 30 °C.

Many people predicted that the market for mycoprotein for human consumption would be very uncertain because there was no guarantee that consumers would find the product acceptable. One of the economic factors that encouraged RHM to develop this product was that waste products from food processing plants could be used to provide the sugar supply. The research and development programme took over twenty years to bring the product to the market (see Table 7.4.1). The mycoprotein production plant had to be carefully designed to allow for two key features. The process would have to be continuous to avoid interruptions to production and it should maintain sterile conditions to prevent contamination by pathogenic microbes.

Table 7.4.1 Mycoprotein research and development programme.

1964	Product development of mycoprotein begins. Animal trials begin to test toxicity.
1969	Start-up of simple pilot production plant.
1973	Development of scaled-up production plant using continuous processing.
1974	Trials on humans begin, testing the acceptability of the product to the consumer.
1976	Development of techniques to convert raw mycoprotein into an attractive food ingredient.
1978	Submission to the government to gain approval for sale to the public.
1981	Government gives preliminary clearance for marketing mycoprotein for public consumption.
1985	Launch of the first mycoprotein product in the shops, a Quorn savoury pie.
1987	Launch of the first Quorn recipe dish in British Home Stores restaurants.
1990	Launch of Quorn as an ingredient for home-cooking (London area only).
1991	March – Quorn output had reached 22 tonnes per week.

ACTIVITY

Marketing mycoprotein

1. Use the information in Table 7.4.2 to explain why a burger made from mycoprotein is healthier than one made with beef.

2. Devise a test to see if people can taste the difference between a dish made from mycoprotein and the same dish made from meat.

3. Figure 7.4.1 shows the annual worldwide sales of mycoprotein since 1985. Account for the trend it shows. Do you think this trend is likely to continue?

4. Find out the range of foods that contain Quorn and try tasting some.

5. Consumer testing of Quorn was more important than for most other new food products. Why do you think this was?

6. Explain why animal trials preceded human trials of mycoprotein.

7. Suggest why a pilot production plant was set up at an early stage in the research and development programme.

8. For what sorts of questions would the Government need answers before it gave clearance for the public sale of mycoprotein?

Table 7.4.2 Nutritional analysis of mycoprotein and beef.

	Percentage content (dry mass)	
	Mycoprotein	Lean beef steak
Protein	47	68
Fat	14	30
Fibre	25	0
Carbohydrate	10	0

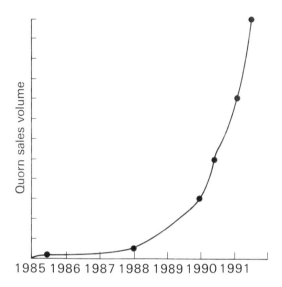

Figure 7.4.1 The increase in sales of the mycoprotein Quorn® from 1985 to 1991.

7 FEED THE WORLD

UNIT 7.5 Intensive agriculture and the environment

Agriculture was one of the earliest human inventions to have a widespread effect on the natural environment. In this unit, we look at the way in which modern agricultural technology can have a harmful effect on ecosystems.

WHAT IS AN ECOSYSTEM?

An ecosystem consists of all the organisms living in a particular location. A local pond or the underside of a brick lying on soil are examples of natural ecosystems that you can readily investigate. Many natural ecosystems have been polluted by the waste products of the most technologically advanced species on the planet, *Homo sapiens*, by agriculture as much as industry. We have also deliberately changed or destroyed natural ecosystems to set up artificial ecosystems in which specific plant and animal species are harvested for food. Greenhouses, oyster beds and trout farms are examples of carefully designed and controlled environments that will provide the optimum conditions for the growth of species that we need for food.

Monoculture is the most widespread system of artificial plant ecosystem on the planet. The ground is cleared of the huge variety of species that makes up the natural ecosystem and replaced by fields where a single crop species such as wheat, rice or maize is grown. Such an artificial ecosystem is very difficult to sustain. Other species that migrate in to colonise it must be removed by herbicides (weedkillers) and pesticides. Because the individual crop plants are genetically identical they are very prone to attack by disease. Heavy applications of chemical fertilisers are needed to replace the nutrients removed from the soil by the densely planted crops.

ENVIRONMENTAL EFFECTS OF MONOCULTURE

Intensive agriculture has certain undesirable effects on the environment. Pesticides sprayed on to crops may be washed into lakes and pass up through food chains, accumulating in dangerous concentrations in the organisms at the top. Nitrates and phosphates from fertilisers are washed from the soil into rivers and lakes, causing **eutrophication**. This occurs when a thick layer of algae forms on the water surface, fed by the fertiliser, which encourages the growth of bacteria that feed on it. These bacteria use up the oxygen in the water, preventing the survival of any other life. Such lakes and rivers are 'dead'.

Nitrates, phosphates and pesticides washed into lakes also enter humans in very small quantities through tap water. The long-term effects of low concentrations of these chemicals on our health are unclear. A link with cancer has been suggested, but this is very difficult to prove because of the impossibility of separating out the different influences that lead to the onset of this disease.

Resistant pests pose a worldwide danger

Scientists, industry and governments must stop the world rushing back to the dark ages that existed before the discovery of insecticides and antibiotics. 'The irresponsible use of insecticides and antibiotics is producing strains of monster bugs that are immune to our chemical weapons' warned Robert Metcalf, Professor of Entomology at the University of Illinois.

The only way to slow the process is to work out a coordinated policy for controlling pests and pathogens, and to impose controls on the users of these chemicals and the industries that produce them.

More than 500 species of insect are immune to insecticides. About 10 to 15 are immune to just about everything and are very hard to kill. Likewise, many bacteria are resistant to a range of antibiotics and can pass on their resistance to unrelated bacteria in addition to their own offspring.

The development of a new insecticide is becoming prohibitively expensive. Nowadays scientists may have to look at 20 000 chemicals before finding a single useful one. The average cost of research and development before a successful new pesticide reaches the market is $50 million. As a result some companies have given up trying.

In the case of antibiotics, people's lives are in danger. Some bacteria have developed immunity to the entire spectrum of common antibiotics. The excessive use of antibiotics in both medicine and agriculture is to blame. Half the antibiotics produced in the USA are given to farm animals, sometimes as a protective measure and sometimes to make them grow faster. In some countries where controls are not very stringent, farmers spray antibiotics directly on to fruit trees to beat bacterial infections. The more bacteria are exposed to the drugs, the faster they develop resistance.

Scientists are recommending a number of ways to combat the menace of insecticide and antibiotics abuse. An approach known as 'integrated pest management' involves the limited use of pesticides alongside other means of control. Resistant species are more likely to succumb to attack by a range of weapons.

Rotation of crops, the breeding of pest-resistant strains and biological control by the pests' natural predators should be on the list of options. If farmers apply insecticides only as a last resort, resistance will develop at a much slower rate. Insecticides that attack only one stage of the insects' life cycle are another useful option. If the larva develops resistance there is always the chance of attacking the adult.

Some scientists believe that the best defence is to breed resistant strains of plants that are immune to attack by pests. However there are still considerable problems to be overcome. The majority of farmers prefer chemicals because they are often cheaper than resistant plant strains. They are backed up by the agrochemical companies, who are worried by a drop in sales of their chemical pesticides. John Barrett of Cambridge University is committed to the pest-resistant plants approach. He accuses the agrochemical industry of trying to discredit the work of plant-breeders: 'It tries to convince people that chemicals are the only solution. Attempts to introduce sensible control programmes are in danger of being stillborn because of the shorter term interests of the political and financial lobbies.'

The global nature of the problem also causes complications. Resistant bacteria do not need passports and they can move anywhere. Whatever strategies for control are planned, they must be carried out internationally.

Based on information in the *New Scientist*, 26 February 1987.

ACTIVITY

Coping with pests

Read the article before answering these questions.

1. Why does Professor Metcalf describe the use of insecticides and antibiotics as 'irresponsible'?

2. Why are some companies giving up trying to develop new chemical pesticides?

3. A particular pest species tends to specialise in feeding on one specific plant species. Crop rotation involves sowing several different crops in successive years in a three- or four-year cycle. How could this reduce the populations of insect pests that feed on the crops?

4. The article mentions two other alternative methods of pest control to replace chemical pesticides. Explain how they work.

5. The article suggests that there are a number of factors slowing down the introduction of integrated pest management. What are these factors? How significant do you think they are?

6. Some people have suggested that farmers should convert to 'organic' methods, which do not use chemical fertilisers and pesticides. Survey your local shops to find out the cost and availability of organic foods. Why is so little farming organic?

ACTIVITY

Biological applications

These questions require a deeper biological knowledge.

1. Using the ideas of gene mutation and natural selection, give a detailed explanation of the evolution of insecticide resistance in pests.

2. The passing on of resistance from one bacterial species to another is achieved through a mechanism of genetic transfer referred to as 'conjugation'. Find out the details of this mechanism.

Science, Technology and Society © D. Andrews, 1992

7 FEED THE WORLD

UNIT 7.6 The Green Revolution

In the 1960s the United Nations set out to tackle the problem of famine by launching a worldwide campaign to bring about the 'Green Revolution'. It was intended to apply the latest technology to encourage radical changes in the agricultural and food processing systems of the Third World and so increase their output of food. In some respects this was a considerable success. The world's food production per head of population has increased. But as you can see in Table 7.6.1, this improvement has not been uniform across continents.

This unit examines some of the reasons for the variation in results in different parts of the world.

In the early stages of the revolution many mistakes were made, which are now being appreciated and remedied.

- Irrigation schemes were used to bring previously infertile land into cultivation. In many cases the high levels of evaporation led to salinisation of the soil. The salt levels were so high that crops could not be grown.
- Deforestation and overgrazing of the land by livestock increased the erosion of soil by wind and rain. In arid regions previously fertile soils were turned into desert (desertification).
- The heavy use of pesticides and fertilisers initially increased crop yields but caused long-term pollution problems (see Unit 7.5).
- New seed varieties bred to resist disease and produce increased yields were introduced. Unfortunately they were only viable on well-irrigated soils that received heavy applications of fertiliser. They were of little use to the poorer farmer.

It became apparent that solutions relying on advanced Western technology could not be transferred wholesale to the Third World. Very often experts from the developed countries could not appreciate the local cultural and economic conditions that hampered the introduction of new technology. The case studies illustrate this problem. According to the influential economist, Ernst Schumacher, the best approach is to encourage small-scale solutions that emerge through self-help. He is famous for the following quote, that summarises this philosophy:

> 'Give a man a fish, as the saying goes, and you are helping him a little bit for a very short while; teach him the art of fishing, and he can help himself all his life ... but teach him to make his own fishing tackle and you have helped him to become not only self supporting but also self reliant and independent.'

ACTIVITY

1. For each of the case studies suggest how the problem illustrated could have been avoided.

2. Use the information in Table 7.6.1 to describe the worldwide variation in food production between continents.

3. Is there evidence from Table 7.6.1 that fertiliser use is responsible for the increase in food production?

Table 7.6.1 Changes in world food production.

Region	Per capita food production (1961–64 = 100)		Per hectare fertiliser use (kilograms)	
	1961–64	1981–84	1961–64	1981–84
World	100	112	29.3	85.3
North America	100	121	47.3	93.2
Western Europe	100	131	124.4	224.3
Africa	100	88	1.8	9.7
Latin America	100	108	1.6	32.4
Asian CPEs*	100	135	15.8	170.3
Far East	100	116	6.4	45.8

*Centrally planned economies e.g. China

Case studies

- In Indonesia loans were provided so that farmers could buy their own tractors. The cost of spare parts is so high that the farmers cannot afford to keep the tractors running when they break down. Abandoned tractors rapidly turning into rusty wrecks now litter the landscape.

- Metal workers in Nigeria use cheap, poor quality material to make low cost machines for grating cassava – a root crop grown as a staple food in many parts of the world. The unreliability of these machines led aid agencies to develop training schemes to teach them how to build sturdier machines with better material.

    After the course some of the metal workers returned to making the machines as they had always done. The customers did not want to pay the extra for an improved machine. The workers preferred this situation because of the regular income they received from repairing machines as they broke down.

- An aid agency set up a project in Papua New Guinea to establish a small-scale baking industry. They trained local blacksmiths to make special fuel-efficient bread ovens and ran courses for women's groups to show how they could use the ovens to generate a steady income. Sixty ovens were produced within 18 months. A few years later a development worker was disappointed to discover that not one of the ovens was in commercial use. The women had abandoned the bakery ovens as soon as they had raised enough money to start up projects that really interested them, such as dressmaking, chicken farms and tradestores.

- In many developing countries mothers have been encouraged to use canned milk powder to feed their children. This has caused increases in cases of gastro-enteritis because often the water used to make up the milk is contaminated with bacteria.

8 MEDICINE, HEALTH AND SOCIETY

UNIT 8.1 Health care: curative and preventative approaches

Much medical research is directed towards finding effective means of diagnosing and treating disease. This **curative** approach has been applied with considerable success to many major diseases. For example, the development of drugs and radiotherapy have both contributed to the greatly improved survival chances of cancer patients.

Medical science has also made great progress in identifying the causes of ill health. It makes sense to use this knowledge to prevent disease rather than waiting to cure people once they become ill. For some diseases for which there is currently no cure, AIDS for example, a **preventative** approach is clearly the ideal. Preventing illness can also be less expensive than treating it with costly surgery or drugs. Preventative health measures in the nineteenth century, such as the provision of clean drinking water, were successful in reducing the death rates from many infectious diseases long before the introduction of antibiotics or vaccines. The high death rates from heart disease in rich countries today also need to be tackled by preventative measures.

The main elements of a preventative programme of health care are:

- A health education programme to encourage people to adopt a healthier life-style.
- Investment in measures to improve the environment, for example, reducing air pollution.
- A well-organised system of public hygiene measures to avoid the transmission of infectious disease, for example, a system of rubbish disposal, sewage treatment plants and the supply of clean drinking water.
- A coordinated attack on the main organisms that transmit disease (vector organisms), for example, the eradication of the Anopheles mosquito, the vector of the malarial parasite.
- Government intervention to tackle poverty and the health risks that result from it, for example, to improve nutrition so that resistance to infectious disease is strengthened.
- A screening programme that regularly checks people for signs of the early stages of an illness, for example, cervical cancer.

THE NATIONAL HEALTH SERVICE

In the UK at the beginning of the twentieth century, health care was a luxury not everyone could afford. Poor people often went without proper treatment when they became ill, relying instead on useless and sometimes dangerous home remedies. The lucky ones received treatment from doctors who gave their services free of charge to their poorest patients. Because people had to pay for treatment, the best hospitals and doctors tended to be in areas of greatest prosperity rather than greatest need.

The injustice of this situation was highlighted by Sir William Beveridge's report in 1941, which called on the government to set up a nationally coordinated system to provide health care for everyone, regardless of their ability to pay. The National Health Service was set up in 1945 on the principle that patients should be entitled to medical services free at the point of need.

In the 1990s the preventative health care offered by the NHS continues to expand. General practitioners offer medical check-ups to give an early warning of potentially dangerous conditions.

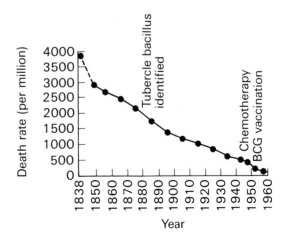

Figure 8.1.1 The change in the death rate for tuberculosis (1838–1960).

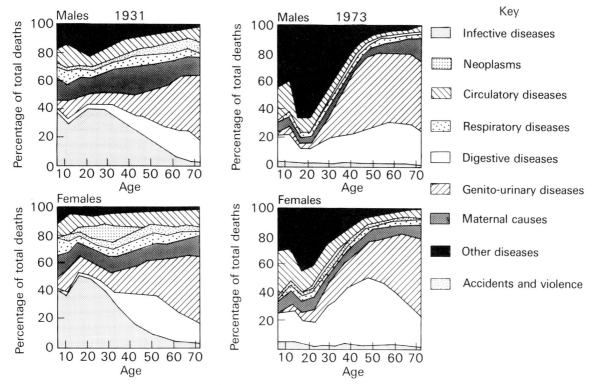

Figure 8.1.2 The causes of death for different age groups in 1931 and 1973.

ACTIVITY

Changing patterns of disease

1. Use the information in Figure 8.1.1 to decide what have been the major preventative health measures in achieving a reduced death rate from tuberculosis.

2. What does Figure 8.8.2 indicate about the differences in the causes of death among teenagers, the middle-aged and the elderly in 1973?

3. Use the information in Figure 8.1.2 to write a short report describing the changes in the major causes of death between 1931 and 1973. Suggest explanations for these changes.

4. Carry out a research project on a particular illness that interests you. Include what is understood about its cause, the problems experienced by its sufferers, and the treatments available. You could also examine the roles of the different types of health professionals who are involved in the diagnosis and treatment of the illness. The following are suggested: AIDS (see Unit 8.3), agoraphobia, Alzheimer's disease, autism, cystic fibrosis (see Unit 5.4), multiple sclerosis, Parkinson's disease (see Unit 8.6), schizophrenia, sickle cell anaemia (see Unit 5.3).

5. In spite of many health education campaigns, diseases linked to smoking, alcohol abuse and poor diet continue to be widespread causes of death. Suggest possible reasons for this. Don't people take any notice of health education?

6. If the NHS were given more money, in which areas of health care should it be spent?

Science, Technology and Society © D. Andrews, 1992

8 MEDICINE, HEALTH AND SOCIETY

UNIT 8.2 Appropriate health care: barefoot doctors

Developing countries are in desperate need of preventative health care. Large hospitals with the latest in high technology equipment, such as those in cities in the industrialised world, are clearly inappropriate to provide such care. In countries with a widely dispersed rural population most people will be a long distance from a major town and their access to medical help would be severely limited if health services were concentrated in this way. Also, there is often insufficient money to train specialist doctors. An alternative approach that directly tackles the particular health needs of the impoverished and poorly educated rural population is required.

In June 1965, the Chinese leader, Mao Zedong, launched an innovatory approach to the country's health care that proved so successful that it was adopted in many other parts of the developing world. Health care was delivered through huge numbers of health auxiliaries, who were chosen by their fellow villagers and trained on a three-month crash course in the essentials of preventative medicine. These part-time health workers spent the rest of their time working barefoot in the paddy fields, like the other villagers, and so became known as 'barefoot doctors'. The following extract describes the work of Victor Charca, a Peruvian barefoot doctor in the remote lakeside village of Machajmarca, in the Andes.

ACTIVITY

Victor Charca

Read the extract about Victor Charca before answering these questions.

1 In what ways is a health care system based on barefoot doctors more appropriate to the needs of a developing country than one based on a comparatively small number of highly trained doctors?

2 What criticisms might qualified doctors have of barefoot doctors?

3 What assistance should be provided to ensure that the barefoot doctor can make accurate diagnoses?

4 What are the benefits of the barefoot doctor being chosen from among the villagers rather than being an outsider?

5 Discuss the following statements made by a US health worker after many years' experience working with barefoot doctors (village health workers) in Latin America.

- 'The health of the people is far more influenced by politics and power groups, by distribution of land and wealth, than by the treatment or prevention of disease.'
- 'The village health worker helps to liberate the community not only from outside exploitation and oppression but from their own short-sightedness and greed.'
- 'The day must come when we look at a village health worker as the key member of the health team and the doctor as the auxiliary... so the doctor is on *tap*, not on *top*.'

Victor Charca – barefoot doctor

Charca was elected at a village meeting of the ninety-two families of Machajmarca. He spends about two hours a day on health work, but receives no salary. Instead, his services are paid for by his patients. The community itself met to decide an appropriate scale of fees and charges for drugs, and built with its own hands the smart, mauve-painted, tin-roofed building on a hill that serves as Charca's health post. Inside the long, one-roomed building, wooden benches line the walls for patients to wait on.

Charca has made a handwritten poster for his patients' guidance, which lists the services offered by the health auxiliary: 'Sanitary education, home visits, first aid, tuberculosis examinations, vaccinations, attendance at births, supply of medicines, personal medical attention, coordination with local authorities and institutions, daily and monthly reports on activities.' In fact, his functions are broader than the brief catalogue suggests. He provides attention to mother and child, with pre- and post-natal care and advice, as well as acting as the midwife at the birth. He gives vaccinations against tuberculosis, measles, polio, smallpox, yellow fever and rabies, and a triple dose against whooping cough, tetanus and diphtheria. It is his responsibility, also, to organize vaccination campaigns to inform and convince people of the need for immunization. He promotes sanitary improvements such as latrines or water filters, and tries to teach his community, through meetings and talks, to prevent disease by good hygienic practices. As he goes along, he keeps a continuous census of births, illnesses and deaths with their causes. And he is, of course, trained and equipped to deal with the most common and straightforward ailments such as diarrhoea, pneumonia, aches and sprains, wounds, eye infections, stomach pains, anaemia and skin complaints.

Charca's technology is basic, boiled down to the bare essentials needed to do the job. His transport is a bike, provided by UNICEF. His auxiliary's kit prepares him for most eventualities, consisting of a limited number of basic drugs such as analgesics (in pill-syrup and injectable form), a few antibiotics and sulpha drugs, eye ointment, chest rubs, deworming pills and powders, along with cotton, gauze, bandages, hypodermic syringes and needles, scissors, thermometers, a sterilizing dish and two planks of wood to make a stretcher. It is up to him to keep the kit fully stocked, covering costs with the fees charged, and he is free to add extra medicines if he finds them useful.

from *The Third World Tomorrow*, by Paul Harrison

Reprinted with the permission of the Peters Fraser & Dunlop Group Ltd.

8 MEDICINE, HEALTH AND SOCIETY

UNIT 8.3 AIDS/HIV campaigns

AIDS is a global disease. The World Health Organisation estimates that by the year 2000 about 6 million people worldwide will have contracted AIDS. Every country in the world is having to come to terms with the fact that we are faced with an infectious disease for which there is currently no cure or vaccine. At present the only way to stop it spreading is to take preventative action to block the routes through which the virus can be passed from one person to another (see the box).

The richer countries of the developed world have a huge range of resources that they can draw upon to tackle the AIDS crisis. The poorer countries have particular difficulties to overcome. They lack the resources to mount an extensive programme of education to inform people how they can combat the spread of HIV.

AIDS IN INDIA

In poor countries, such as India, the desperate measures people take to avoid poverty and hunger contribute to the spread of HIV infection. Professional blood donors are an accepted part of the health service in India. For donating a pint of blood they are paid about 35 rupees, the equivalent of a day's wage. In cities such as Bombay, blood donors are now screened for HIV and the risk of becoming infected through a transfusion is minimised. However, in many of the rural areas of the country they lack the facilities to check for the presence of HIV in donated blood.

In the cities, with the ever-present threat of hunger and homelessness, large numbers of women resort to prostitution as a means of survival. The prostitute population of Bombay is estimated to be around 100 000 and about one third of these are thought to be HIV positive. Health workers have great difficulties in trying to get over the message about the risk of AIDS to this group. 'They are in a buyer's market and are not in a position to insist on their clients using condoms. They do not want to listen to us when all they are worried about is where the next meal is coming from.' The large numbers of migrant male workers who spend months or years away from their homes and families, add to the problem.

How is HIV transmitted?

A person who is infected with Human Immune Deficiency Virus is described as being HIV positive. In the early stages of infection an HIV positive person will suffer no symptoms and notice no change in their health. Unless they go for a blood test they will be unaware that they are HIV positive, and could pass on the virus to someone else.

HIV infection will eventually cause a serious weakening of the ability of the white blood cells to fight off pathogens. In most cases this leads to the development of several life-threatening illnesses, including a rare form of skin cancer and certain serious lung infections such as pneumonia. When the symptoms of these illnesses appear the person who is HIV positive has now developed AIDS (Acquired Immune Deficiency Syndrome). At present it appears that, once infected with HIV, a person can expect to live, on average, for a further eight to nine years. However about 10 per cent of people who are HIV positive will not have developed AIDS ten years after infection.

The first cases of AIDS in the UK were in male homosexuals. However, the rate of infection among gay men in the UK is now falling, while the rate of infection among heterosexual men and women is rising.

- It is very difficult to transmit the virus from one person to another. The virus is present in the blood, semen and vaginal secretions. The only way that you can become infected is if one of these fluids passes into your bloodstream.

- HIV can sometimes be transmitted from an infected mother to her child during pregnancy, birth or breastfeeding (HIV is present in breastmilk).

- HIV is destroyed outside the body by the following:
  a) when the cell in which it is carried dries out
  b) applying bleach
  c) boiling in water.

ACTIVITY

1. Use the information in the box to decide whether or not the following activities could transmit HIV:
   a) sharing cups and cutlery
   b) sharing needles for injecting drugs
   c) shaking hands
   d) kissing
   e) swimming
   f) sexual intercourse
   g) blood transfusions
   h) blood from an infected person's wound entering a cut.

2. How would high levels of illiteracy hamper the campaign to control the spread of HIV infection?

3. TV and radio advertising have been important elements in the AIDS campaign in rich countries. How important are they to similar campaigns in poorer countries such as India?

4. What methods would be most suitable for an AIDS campaign in
   a) poor countries
   b) rich countries?

5. How does the failure to screen donated blood for HIV affect the spread of the disease?

6. Discuss the following statements.
   - 'Keeping sex within marriage protects you from HIV and AIDS.'
   - 'It serves you right if you get AIDS.'
   - 'I don't need to worry about AIDS: I'm not gay and I don't inject drugs.'

7. A haemophiliac schoolboy has been found to be HIV positive owing to a transfusion of unscreened blood. Some parents, staff and pupils say he should not be allowed to attend school.
   - What is your attitude towards this situation?
   - What would you do if you were:
     a) the headteacher
     b) a parent?

8. Design and conduct a survey to assess young people's attitudes to AIDS and HIV infection. You could start by providing a list of statements like those in question 6 and asking whether people agreed or disagreed with them. Make sure you follow up your survey with a test of students' knowledge about AIDS/HIV so that any misunderstandings or false beliefs can be remedied.

   How would you analyse your results to find out whether people's level of knowledge of AIDS/HIV has a significant influence on their attitude to the disease?

9. If you wish to find out more about AIDS and the transmission of HIV, write to:
   Terrence Higgins Trust,
   BM/AIDS,
   London WC1N 3XX.

UNIT 8.4 Health and social class

Our state of health is influenced by many aspects of our environment. For example, damp housing conditions and pollution in the air we breathe can increase the risk of lung disease. Certain aspects of our life-style may also carry a health risk: our diet, smoking behaviour and the amount we exercise will affect our chances of having a heart attack. Maintaining good health also relies on taking full advantage of the health services that are available to us.

All these factors are related to social class and they operate to create a serious gap between the health of the richest and poorest sectors of our society.

SOCIAL CLASS

There are several ways of classifying people into social classes. One of the most common approaches is to use the person's occupation. The system in Table 8.4.1 was developed by the Registrar-General and is widely used in surveys carried out by the government and the health service. Although the occupation's status is the sole criterion used in this classification system, income is also closely correlated with social class.

Table 8.4.1 The five occupational classes.

Occupational class	Examples of occupations
1 Professional	Lawyer, doctor, chartered accountant, university lecturer
2 Managerial and technical	Schoolteacher, librarian, nurse
3 (Non-manual) Clerical and supervisory	Bank clerk, sales representative
3 (Manual) Skilled workers	Mechanic, bricklayer, cook
4 Semi-skilled	Bus conductor, postman, fisherman
5 Unskilled	Cleaner, labourer, porter

ACTIVITY

Patterns of disease

Studying the death rates for specific causes gives us an insight into the pattern of health problems experienced by a particular group of people. Table 8.4.2 compares the death rates for the whole of England and Wales with the death rates in two Area Health Authorities in very different parts of London. Bromley is a relatively wealthy suburb of London, neighbouring the Kent countryside. City and East London includes the poorer inner city Boroughs of Hackney and Tower Hamlets.

Study the table and then answer the following questions.

1. What causes of death are:
   a) significantly higher
   b) significantly lower

   in the inner city in comparison with the national average? For each cause of death suggest possible explanations for these differences.

2. Describe the main differences between the pattern of mortality in Bromley and in England and Wales as a whole. What social and geographical characteristics of Bromley might explain these differences?

3. Is there any evidence from the table to indicate that air pollution is more serious in East London than in Bromley?

4. For England and Wales as a whole, what proportion of deaths are due to cancer and heart disease combined? How would you expect this figure to compare with figures from 100 years ago?

Table 8.4.2 Death rates (per 100 000 of the population) classified by cause of death.

Cause of death	City and East London	Bromley	England and Wales
Lung cancer	99	70	69
Breast cancer	25	28	24
All cancers	291	267	254
Diabetes	13	8	10
Heart disease	356	350	390
Pneumonia	116	109	105
Bronchitis and emphysema	56	35	45
Peptic ulcer	13	10	8
Accidents	30	20	23
Suicides	12	10	12
Cirrhosis of the liver	6	3	4
All causes	1176	1063	1169

ACTIVITY

Infant mortality and occupational class

Study the data in Table 8.4.3 before answering these questions.

1. a) What is the trend in infant mortality from occupational class 1 to class 5?
   b) What differences in social conditions between the classes might explain this trend?

2. a) Suggest possible causes of babies being born with a below-average birthweight.
   b) What is the trend in babies' birthweight from class 1 to class 5?
   c) What differences in social/economic conditions might be responsible for this trend?

3. Find out what type of help and advice is available through antenatal clinics.

4. Suggest why it is advisable for women to make their first visit to the antenatal clinic at an early stage in pregnancy.

5. How is the timing of a pregnant woman's first visit to the antenatal clinic related to her occupational class? What might be the underlying reason for this link?

6. Using only the information available from the table, suggest how differences in access to the health services might explain the class differences in infant mortality. Draw suitable graphs to illustrate this hypothesis. What other data would you need to collect to investigate this hypothesis?

Table 8.4.3 Infant mortality rates by occupational class.

Occupational class	Infant mortality (Deaths per thousand live births)	% of women whose first visit to antenatal clinic is after 20 weeks of pregnancy	% of babies with birthweight less than 2.5 kg
1	13	27	4.5
2	15	30	4.5
3	18	31	5.6
4	22	35	8.2
5	35	41	8.2

8 MEDICINE, HEALTH AND SOCIETY

UNIT 8.5 Attitudes to abortion

The termination of a pregnancy by artificial means – an abortion – was illegal in most countries until the early 1970s. The methods used to carry out illegal abortions were often dangerous and sometimes caused infertility or death. Nowadays there is a range of safe techniques available for performing abortions. Although advances in medical technology have minimised the risk to women's health, the decision to have an abortion is usually complicated by mixed feelings. On the one hand there may be strong reasons for not having the child, while on the other there may be a strong reluctance to end the life of a developing fetus. Technology has provided the option of a safe abortion but it is up to us to decide the circumstances in which it should be used.

ACTIVITY

Reasons for abortion

In each of the situations described in Box 1 an abortion might be considered the best course of action.

1 Add to this list some suggestions of your own of situations where an abortion might be considered.

2 For each of these situations discuss:
- The likely consequences of the pregnancy continuing and the child being born.
- The likely impact on the situation if the abortion is carried out.
- The likelihood that a doctor in the UK would agree to an abortion under the legal requirements of the Abortion Act (see Box 2).
- Your own views on the acceptability of an abortion in each case.

1 Situations that may justify an abortion

1 The child would be born severely mentally handicapped and would need constant care and attention for the rest of its life (estimated to be 40 years).

2 The child would be born severely physically disabled but with normal intelligence and lifespan.

3 The child will be born with an inherited disorder and although it will be healthy during its early childhood, it is likely to die before the age of twelve.

4 The pregnancy was due to the failure of a sterilisation operation.

5 The pregnancy was due to the failure of the couple's method of contraception.

6 The couple already have two children and feel that they could not afford the cost of another child.

7 The mother is suffering from an acute mental illness, such as depression.

8 The pregnancy is the result of a rape attack on the woman.

9 The couple's relationship is going through a difficult patch and the strain of a baby is likely to end it.

10 The woman is sixteen and has no knowledge of the child's father's whereabouts.

11 The continuation of the pregnancy poses a risk to the mother's life.

12 The fetus is found to be female and the mother feels considerable pressure from the expectations of her culture to have a son.

② The 1967 Abortion Act

Under this Act an abortion may be legally carried out if two doctors agree that there are the following specific grounds for it.

1 That the continuance of the pregnancy would involve:
   - risk to the life of the pregnant woman
   - or risk of injury to her physical or mental health
   - or risk of injury to the physical or mental health of her existing children.

2 That there is a substantial risk that if the child were born it would suffer serious mental or physical handicap.

Under the 1967 Act it was illegal for a doctor to carry out an abortion if the woman's pregnancy had lasted more than 28 weeks.

However, rapid advances in the medical technology for saving the life of premature babies began to raise serious dilemmas for the doctors. With the use of sophisticated incubators, babies born at 24 weeks' gestation are now capable of surviving. In 1989 government legislation was passed to reduce the time limit for abortion to 24 weeks, except for cases of severe physical or mental handicap.

③
- Abortion is the equivalent of murdering a child.
- The fertilised egg has the same right to life as a newborn baby.
- A woman has the right to control her own body and that includes her right to decide whether she needs an abortion.
- Abortion is too easily available.
- A fetus does not gain the rights of a human being until it is old enough to survive outside the uterus.

④
1 Strongly disagree with the statement
2 Mildly disagree with the statement
3 Neither agree nor disagree with the statement
4 Mildly agree with the statement
5 Strongly agree with the statement

ACTIVITY

Abortion attitude survey

People often have strong views and beliefs on abortion. The measurement of attitudes is a complex area, which you can begin to investigate using the following approach.

1 Devise a set of statements on abortion such as those shown in Box 3.

2 Prepare a record sheet to record how strongly people agree or disagree with each statement, using a five-point scale such as that shown in Box 4.

3 For each person you interview, ask additional questions to gain relevant personal information that you think might be linked to their views on abortion. For example, you could ask, 'Do you regularly attend religious services?'

4 Analyse the data from your survey to see whether any patterns emerge in the attitudes that people hold.

Science, Technology and Society © D. Andrews, 1992

SCIENCE TECHNOLOGY AND SOCIETY

8 MEDICINE, HEALTH AND SOCIETY

UNIT 8.6 Medical use of fetal tissue

Should cells from an aborted fetus be used as part of a medical treatment? This is the question that hit the headlines in March 1988 when doctors in Birmingham carried out an operation that was the first of its kind in the UK. A sample of cells was removed from an aborted fetus and implanted in the brains of two middle-aged patients suffering from Parkinson's disease (see Box 1). The publicity that this radical new treatment received generated much controversy and raised a number of issues, which are considered in this unit.

Although some politicians called for a ban on the technique, this pressure was resisted by the Government. It was decided to allow the medical profession, through their own professional body, the British Medical Association (BMA), to provide guidelines to regulate the use of fetal tissue.

---

① Parkinson's disease

Parkinson's disease is a distressing and at present incurable condition that can strike young and old alike, although it is more common in the elderly. It is estimated to affect 100 000 people in the UK and about 1 per cent of the over-65 age group are sufferers.

In the disease cells in a part of the brain called the *substantia nigra* (see Figure 8.6.1) begin to self-destruct for no apparent reason. These cells normally produce the substance **dopamine**, a brain hormone that is vital for coordinating voluntary muscle movements. Without sufficient dopamine, sufferers of Parkinson's disease find that their movements are seriously disrupted. Their muscles stiffen, and actions that we take for granted such as walking, speaking and swallowing may require a tremendous effort of will. Sometimes the muscles will 'freeze' for several minutes. In some cases people have muscle tremors, which may only affect one limb or one side of the body. Other brain functions, such as memory, problem solving and language comprehension, are unaffected by the disease.

Possible treatments

In the early 1960s, when the link between a lack of dopamine and Parkinson-type symptoms was first discovered, the possibility of a successful drug treatment was raised. A drug called L-dopa, which is converted to dopamine in the brain, was very successful in alleviating the symptoms in certain patients, giving them greatly improved control of their movements. However it did not prove to be a long-term solution. After several years on L-dopa, many patients began to suffer from side effects, such as extreme restlessness, that were almost as distressing as the disease itself.

Researchers began to look for an alternative therapy that might avoid the problems of L-dopa. The possibility of using cells from the *substantia nigra* of fetal brains was one hopeful avenue of research. Brain cells of eight-week-old fetuses are still dividing rapidly and will therefore survive removal and implantation into the patient's brain. Older fetuses grow less rapidly and are therefore a less suitable source of brain tissue.

The use of this technique since 1988 has produced varying results. In some cases the short-term improvement in the patients' health has been dramatic. Other patients, whose condition is more serious, are less likely to make a significant improvement after the operation.

ACTIVITY

Your views on fetal tissue implants

First read the information about Parkinson's disease in Box 1 and Sarah Boseley's article from the *Guardian* of 23 April 1988, which is reproduced in Box 2. Then discuss these questions.

1 The newspaper reports referred to these operations as brain cell 'transplants' whereas the medical journals described them as 'implants'. Suggest possible reasons for this difference.

2 Sarah Boseley reports that some people find the idea of using fetal tissue in this sort of operation to be a 'shocking concept'. What is your own reaction to this operation?

3 Is there a difference between the removal of an organ from a young person killed by an accident for a transplant operation, and implanting tissue from an aborted fetus into a patient with Parkinson's disease? Give your reasons.

4 'If society accepts abortion, then it can have no objection to the use of fetal tissues and organs for the benefit of other human beings.' What do you think of this point of view?

5 Why was the age of the fetus relevant for this type of operation?

6 Consider the case of a woman whose husband suffers from Parkinson's disease. She wants to become pregnant so that cells from her fetus can be used in an implant operation to help her husband.
   - What are the BMA's views on the ethics of this decision?
   - Do you agree?

7 Who should take the decision whether or not to develop this new treatment for Parkinson's disease?
   - Should the matter be left entirely to the doctors carrying out the operation?
   - What about the views of the patients?
   - What would be your view if Parliament passed legislation to restrict or ban outright this new medical development?

8 As with the first heart transplants, this operation carries a degree of risk for the patient. Some scientists have argued that its safety and effectiveness should have been fully tested on monkeys before the operation was carried out with humans. What do you think?

9 If you were a doctor involved in developing this new technique, what reasons would you put forward to support the use of fetal brain cells for treating illnesses such as Parkinson's disease?

10 If you were a pregnant woman seeking an abortion, what thoughts and feelings might affect your decision when asked to consent to the use of your aborted fetus for medical research or treatment?
   - What counselling and information would you want?
   - Do you think that the father should have a right to be consulted too?

11 In some hospitals, all women who have an abortion are asked whether they are willing to consent to the use of the fetus for medical research. About 50 per cent give their consent.
   - How might the response be affected if the women knew that some of the fetal cells would be implanted into the brain of another person?
   - Would you criticise a doctor for failing to mention the possibility of a fetal tissue implant when seeking consent from women about to have an abortion?

12 The BMA guidelines rule that doctors involved in fetal tissue implant operations should have nothing to do with seeking consent from women having abortions.
   - Why is this considered to be an important safeguard?

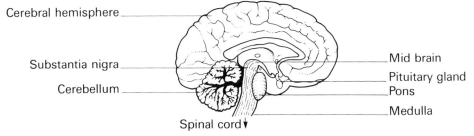

Figure 8.6.1 The substantia nigra is the part of the brain that degenerates in patients with Parkinson's disease.

Transplanting cells from foetuses

Experiments on human embryos

Sarah Boseley

TRANSPLANTING cells from the brain of an aborted foetus into that of a sufferer from crippling Parkinson's disease is a dramatic, and some would say, shocking concept. But in medical terms, science has simply taken the next step forward.

The operations that took place on March 3 and April 7 at the Midland Centre for Neurosurgery and Neurology in Birmingham have already triggered an impassioned ethical debate. The question now is whether this development will lead to new Parliamentary controls.

Foetal material, resulting from both abortions and miscarriages, has been stored and used for research into genetic abnormalities for some years.

Professor Edward Hitchcock has now made history with the first transplants in this country of brain cells, but work on transplanting pancreatic islot cells from foetuses to diabetics is even further advanced. Doctors find the development exciting because it could offer a real cure for the disease. The cells, which would be implanted near the liver, would grow into a clump of tissue which would enable diabetics once more to produce insulin for themselves.

But transplanting cells from foetuses is not suddenly a miracle cure-all. The only diseases that might be treatable at present are those where the patient has a deficiency of a single type of cell which can be surgically implanted – in Parkinson's disease, the cell that produces the chemical dopamine, which helps give us control of our movements.

Professor Hitchcock's operations were carried out on dead foetuses that had been aborted at between six and nine weeks old. They are at that stage only an inch long, and the brain cells have the capability for a short time to continue dividing. In an older foetus it is thought this would not happen quite so rapidly. The cells have to be implanted during major neurosurgery in a deep part of the recipient's brain, the *substantia nigra*, where they will form a colony and produce dopamine.

The women undergoing abortion in Birmingham had consented beforehand to the use of the foetuses for experimental research, although they were not told about the transplant.

Controls on the use of foetal material have been in force since the DHSS issued guidelines after the Peel Report of 1972. Research on a foetus that is viable after separation from the mother is forbidden, and viability is set down as notionally 20 weeks.

While work on foetal material was limited to laboratory inquiries into why abnormal babies are born, there was little fuss. Now, however, alarming images have begun to surface of women becoming pregnant in order to provide a relative with the brain cells they need to be cured of a dreadful disease.

The British Medical Association's ethical committee has anticipated the furore. After six months of discussions, they will consider draft guidelines on the use of foetal material this week. Their final version will be put before the BMA council in May.

Dr Vivien Nathanson, assistant secretary to the committee, stressed the importance of asking the permission of the woman donating the foetus. She said: "The woman should wherever possible be given the choice of saying this is something she agrees with or not." The need for foetal material should not in any way affect the decision whether to abort or the way in which it was done. She added: "It is totally wrong to allow women to become pregnant to become tissue donors." The tissue should be sent anonymously to a separately-run unit, so that doctors performing abortions have nothing to do with cell transplants.

Whatever the BMA rules, however, she is aware that public opinion, now in turmoil over the issue, could sway Parliament to put a stop to the recycling of foetal cells. Following the Warnock Report, a Bill regulating *in vitro* fertilisation will go before this Parliament. In a White Paper, the government has promised a free vote on whether experiments on test-tube embryos, which at the moment do not live beyond 8 or 9 days, should be allowed.

While all the doctors, and the DHSS, insist aborted or miscarried foetuses, which are dead before any transplant begins, are a different issue, Dr Nathanson thinks there may be repercussions. "If Parliament decides because of public pressure that it is unacceptable to experiment on human embryos at less than 8 to 9 days," she says, "there is a much higher chance that they will be under pressure to limit in some way this use of foetal material."

© Sarah Boseley THE GUARDIAN

ACTIVITY

Role play

Explore these issues further by staging interviews in pairs, with one person taking the role of a doctor, the other the role of a young woman about to have an abortion (see Box 3).

After the interviews discuss your reactions to them as a class.

- How did the doctors argue their case?
- How did the manner and attitudes of the doctors affect the decision made by the women?
- What arguments were raised by the women interviewed for and against consent?

3

The doctor
You have to outline the implant procedure, explain why it is important to the patient, and seek the woman's consent for the use of fetal cells to treat someone suffering from Parkinson's disease.

The woman
You are undecided about giving your consent for the use of your aborted fetus for the operation. This is one more thing to worry about at a complicated time in your life, which already seems sufficiently upsetting without this extra problem.

9 WORLD POPULATION – PROBLEMS AND SOLUTIONS

UNIT 9.1 Population growth: the demographic transition

According to the United Nations, the population of the world reached 5000 million on 11 July 1987. It is continuing to increase by over 200 000 people each day. This is equivalent to the population of Birmingham being added to our world every four days. If the rate of growth does not decrease there will be 10 000 million people living on the planet in the year 2010. This growth in numbers places an enormous pressure on the Earth's land and energy resources and poses one of the most intractable problems we face. If we could bring population growth under control it would help developing nations to afford improvements in their economy, education and health services. This unit looks at why this rapid growth has occurred and why it is proving difficult to bring it under control.

For most of our history the size of the world human population has remained steady and it is only relatively recently that rapid population growth has occurred. This growth is largely a result of the dramatic drop in the death rate in many countries. Young children are more likely to survive into adulthood and the average lifespan is getting longer. At the same time the overall birth rate of the world's population remains very high. The combination of the low death rate and the high birth rate means that there is a rapid increase in the world's population. Although there is a wide variation between countries, birth rates are generally coming down very slowly. The wide gap between birth and death rates must be narrowed as soon as possible if we are to bring population growth under control.

THE DEMOGRAPHIC TRANSITION

The demographic transition describes the change that the world's population must go through to slow down population growth. The four stages are shown in Figure 9.1.1. Rich industrialised countries such as the UK, Japan and the USA have reached the final stage of the demographic transition and their populations have stopped growing. However, the bulk of the world's population, who live in the developing countries of Africa, Asia and South America, have still to reach this stage.

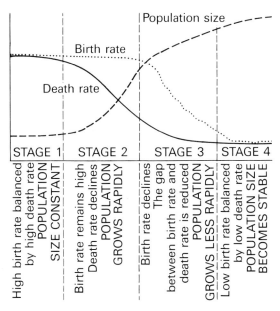

Figure 9.1.1 The four stages of the demographic transition.

Stage 1

The high death rate is a result of famine and disease. Poor sanitation spreads fatal infectious diseases such as cholera and the lack of adequate medical technology reduces the chances of saving lives. Famine occurs when the agricultural technology is not sufficiently advanced to avoid crop failures.

The birth rate is high to offset the large number of infant deaths that occur. This ensures that in each family a sufficient number of children survive to adulthood.

The high birth and death rates cancel each other out so the population size is constant. This has been true for the world's population for most of our history.

Stage 2

The death rate begins to drop due to medical and agricultural improvements. More hygienic sanitation systems, the provision of clean drinking water, and the introduction of mass immunisation and antibiotics contribute to the eradication of infectious diseases and an increase in life expectancy. The likelihood of famine is reduced by advances in agricultural technology, such as the application of fertilisers and improved irrigation techniques.

The birth rate remains high because parents still regard children as an economic asset. This is particularly true in rural communities, where children work on the family farm and look after their parents when they are too old or sick to work.

This increasing gap between birth and death rates causes a rapid increase in the population growth rate – a population explosion.

Stage 3

The birth rate begins to drop. This occurs when people's desired family size reduces owing to economic development and changes in social attitudes. For example, in industrialised urban societies with an emphasis on acquiring skills through extended education, children are no longer an economic asset. As technology introduces more labour-saving devices, women are freed from their traditional domestic role and can seek fulfilment in a career as well as in motherhood. However this change in attitude will not be translated into a reduced family size unless there is access to reliable and socially acceptable methods of contraception (see Unit 9.3).

There seems to be a tendency for changes in birth rate to lag behind changes in death rate partly because of the inbuilt inertia in the high value that societies have traditionally placed on large families.

Stage 4

The birth rate has dropped to a level where it is about the same as the death rate. In practical terms this occurs when the average family size is about 2.1 children. This is described as a stable population with zero growth.

ACTIVITY

Demographic transition

1. Explain how each of the following affect the death rate:
   a) improved sanitation
   b) introduction of mass immunisation
   c) the widespread use of antibiotics.

2. Study the graphs (Figures 9.1.2, 9.1.3, 9.1.4) showing the progress of the demographic transition in three different countries.
   a) For the UK, use the following formula to calculate the rate of population growth (the rate of natural increase) at five-yearly intervals.

   Rate of natural increase = Crude birth rate − Crude death rate.

   b) Plot your calculated data on a graph with time intervals of five years to show the change in the rate of population growth.

   c) Try to work out approximately the years that mark the different stages of the UK's demographic transition.

3. For each country decide which stage of the demographic transition they have reached in 1990.

4. Describe and suggest explanations for the differences in the patterns of change seen in the three countries.

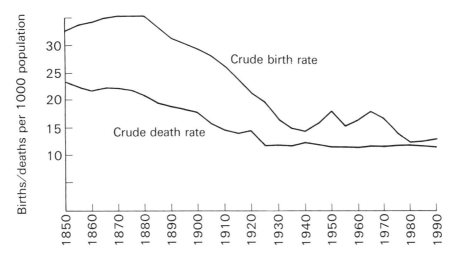

Figure 9.1.2   The annual crude birth and death rates for the UK (1850–1990).

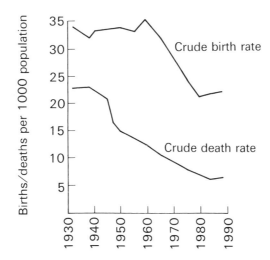

Figure 9.1.3   The annual crude birth and death rates for Chile (1930–1990).

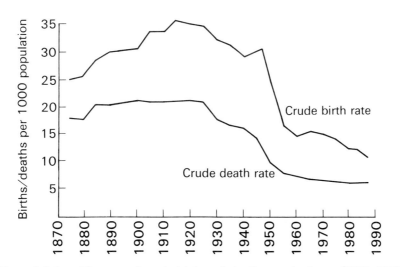

Figure 9.1.4   The annual crude birth and death rates for Japan (1870–1990).

SCIENCE TECHNOLOGY AND SOCIETY

9 WORLD POPULATION – PROBLEMS AND SOLUTIONS

UNIT 9.2 Population pyramids

The numbers in different age groups of a population can have a profound effect on the social and economic life of the country. This is because the adult population must generate the wealth to support the young and the old who are not making a significant contribution to the economy. In most developed countries, the proportion of elderly people relying on a pension for their income is increasing, and will have a major impact on the way society is organised. In developing countries the proportion of young people is very high and this too has its economic effects. A population pyramid is one way in which the different age groups in a population can be displayed. The length of each bar represents the number of people in each sex and age group.

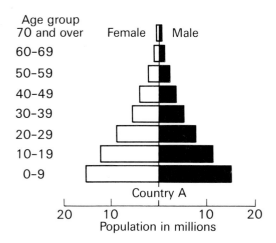

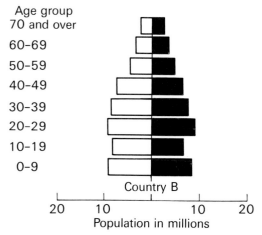

Figure 9.2.1 Population pyramids of two countries.

ACTIVITY

Age structures in developed and developing countries

Study the population pyramids of the two countries in Figure 9.2.1.

1 Which country has the higher death rate? Explain your answer.

2 Calculate the following for both countries:
   a) the total population
   b) the number of people in the 20–59 age group
   c) the number of people in the over 60 age group
   d) the ratio of the number of adults of working age (20–59) to the number of elderly people.

3 a) Which of these two age structures is most like that in a developed country?
   b) In the future, what will happen to the size of the ratio you calculated in question 2d in developed and developing countries?
   c) What is the likely impact of this trend on the economy?

4 Calculate the ratios of economically active adults (20–59 year olds) to economically inactive children (0–9 year olds) for the two countries.

5 Which country will experience economic and social problems caused by a high proportion of children relative to the proportion of adults? What are these problems likely to be?

Science, Technology and Society © D. Andrews, 1992

ACTIVITY

The UK's changing age structure

Figure 9.2.2 shows the age structure of the UK population in 1841, 1891 and 1961.

1. a) What is the main difference between the pyramids in 1841 and 1891?
   b) What has happened to the birth rate during this time?

2. a) How does the size of the population over 60 of 1891 compare with that of 1961?
   b) Suggest possible explanations for the change.

3. a) For the 1961 pyramid, calculate and compare the ratio of males to females in the over-60 and under-16 age groups.
   b) Explain why the two ratios are different.

4. How did World War II (1939–45) influence the population structure?

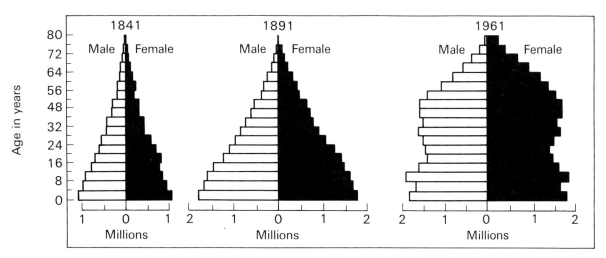

Figure 9.2.2   UK population pyramids at different dates over the last 150 years.

9 WORLD POPULATION – PROBLEMS AND SOLUTIONS

UNIT 9.3 Contraceptive choice

New technology can be invaluable for tackling the many problems that face us – both personal and global. New forms of contraceptive technology are helping individuals to limit their family size as well as contributing to the solution of the global population explosion. However no example of technology is without its drawbacks. This principle is well-illustrated by contraceptive technology, which has produced a vast range of contraceptive methods over the past 100 years.

CONTRACEPTIVE METHODS

You will probably be using some form of contraception for twenty to thirty years of your life. Your choice of method will depend on the personal preferences of you and your partner and your circumstances at the time. These factors will influence your view of the advantages and disadvantages of each one. Developing countries must also make a careful selection of appropriate contraceptive technology if their family planning policies are to be successful. In this unit you will examine various contraceptive methods and attempt to assess their advantages and disadvantages in different situations.

Withdrawal

For most of human history it has been common for women to give birth to more than ten children. Before the twentieth century, the most popular way to stop the conception of a child was the withdrawal method. This is when the man withdraws his penis before he ejaculates so that semen is not released into the woman's vagina. Withdrawal is not a very reliable method and may lead to sexual frustration in both sexes. But it is better than no contraception at all.

Nowadays couples in developed countries have the choice of a wide range of contraceptive methods to help them plan the size of their families and avoid unwanted pregnancies. However in developing countries that lack the finance to develop large-scale family planning services, the use of the withdrawal method is widespread.

Safe period (fertility awareness)

Sperm can survive for up to seven days in the uterus. The egg will die within 24 hours of ovulation if it is not fertilised. By avoiding intercourse on the appropriate days around ovulation the couple can greatly reduce the likelihood of fertilisation. There are two major problems with this method: the difficulty in predicting the exact time of ovulation, especially if the woman's menstrual cycle is irregular, and its relative unreliability compared to other methods. However this 'natural' method of contraception is popular with those who, for personal or religious reasons, object to artificial devices.

Condoms

The first condoms were invented by the Romans who used animal bladders to cover the penis during intercourse. Nowadays condoms are made from very thin latex rubber. They are rolled over the erect penis so that the semen is collected inside. The man must be careful to stop semen leaking out when he withdraws his penis since a single drop of semen can contain up to 3 million sperm! As it is the only method that cuts the risk of spreading sexually transmitted diseases such as AIDS, condom use has increased over the past ten years. The man may feel some loss of sensation although this will vary according to the thickness of the rubber of the brand being used.

No medical advice or assistance is needed to use a condom and they are sold through ordinary stores and vending machines as well as by pharmacists.

Oral contraceptives – the 'pill'

The contraceptive pill was first developed in the 1950s and soon became popular throughout the world. The **combined pill** contains two hormones, progesterone and oestrogen, and is taken by the woman each day for 21 days to stop her from ovulating. She then stops taking it for seven days during which time she has a period. There has been some concern about the higher risk of thrombosis among older pill-users who smoke, but this risk can be reduced by using a pill with a lower oestrogen dose.

Although statistically pill-users are more likely to suffer from cancer of the cervix than non pill-users, this effect may stem from other causes. For example, the higher the number of different sexual partners a woman has, the greater her risk of cervical cancer. However, pill-users have a reduced risk of developing cancer of the ovary or uterus.

A **progesterone-only pill**, the 'mini-pill', is also available. It works by thickening the mucus in the cervix so sperm cannot get through. Although slightly less reliable than the combined pill, it has fewer side effects and health risks.

Before the pill is prescribed a woman must have a health check. If signs of thrombosis or high blood pressure are found the pill must be discontinued.

Intra-Uterine Devices (IUDs)

The IUD is a tiny plastic structure (see Figure 9.3.1) placed in the uterus by a trained medical worker, who inserts it through the cervix using a hollow applicator. It can remain there for up to three years without being replaced and it works by stopping the fertilised egg from implanting in the uterus lining. Some IUDs have fine copper wire coiled around them to increase their effectiveness. The major problem with the IUD is that women who use it have an increased risk of infection of the uterus and fallopian tubes (pelvic inflammatory disease). In some cases this can lead to infertility due to blockage of the fallopian tubes. IUD-users may also suffer from cramps and excessive menstrual bleeding. It is more suitable for women who have had a child, since the uterus is then less likely to reject the IUD.

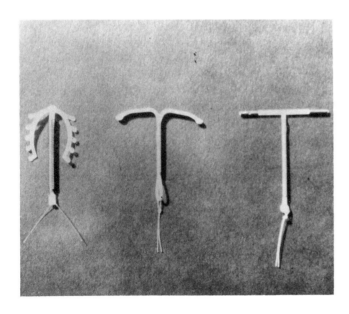

Figure 9.3.1   Intra-uterine devices (IUDs).

The diaphragm

The diaphragm, or cap, is a small dome of rubber with a flexible rim, which the woman covers with spermicide cream and then places in her vagina so that it covers her cervix (see Figure 9.3.2). Sperm cannot penetrate this barrier and are killed by the spermicide. The diaphragm must be left in place for at least six hours after intercourse to ensure that no sperm are allowed to pass through the cervix into the uterus. It has no side effects but some women find it 'messy' to use.

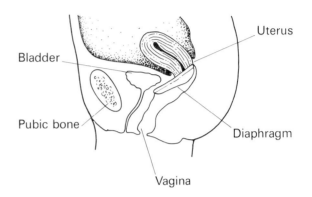

Figure 9.3.2   The diaphragm, when placed over the cervix, acts as a barrier to the sperm, preventing them from entering the uterus.

Sterilisation

Male sterilisation (vasectomy) is a simple operation that can be carried out under a local anaesthetic in less than fifteen minutes. Two small incisions are made in the scrotum so that each of the sperm ducts leading from the testes can be cut and tied. The sperm stay in the testes, where they are broken down and reabsorbed. There is only a small reduction in the amount of semen produced, as most of the fluid is produced by other glands.

Sterilisation for women is more complicated, although in some cases it too can be carried out under local anaesthetic. One type of sterilisation involves cutting and tying the fallopian tubes so that sperm are unable to reach the egg. Sometimes plastic clips are used to close the fallopian tubes. Ovulation and the woman's menstrual cycle are unaffected.

These sterilisation operations provide permanent contraceptive protection and are difficult to reverse. For this reason they are rarely given to people under thirty, unless they have a sound justification for wanting to be sterilised.

CONTRACEPTIVE RELIABILITY

The failure rate of a contraceptive is calculated by the percentage of women using the method who become pregnant during a 12-month period. The failure rate depends on how carefully the method is used. In Table 9.3.1 a range of failure rates is given. The lowest figure shows how effective a method can be if it is used with the greatest care.

Table 9.3.1   Range of failure rates for contraceptive methods.

Contraceptive method	Failure rate
Male sterilisation	0–0.2
Female sterilisation	0–0.5
Combined pill	0.2–1
Progesterone pill	0.3–5
IUD	0.3–4
Condom	2–15
Diaphragm	2–15
Withdrawal	8–17
Fertility awareness	6–25
No contraception (women under 40)	40–45

ACTIVITY

The impact of contraceptives

1. Developing countries often have a widely spread population and a shortage of health service workers.
   Suggest why condoms would be a good method for inclusion in their family planning programmes.

2. The pill was blamed for the increase in promiscuity and in the number of cases of sexually transmitted diseases when it was first introduced in the 1960s. Do you agree?

3. In the UK the use of the contraceptive pill has fallen recently. Suggest possible reasons for this.

4. Suggest one advantage and one disadvantage of relying on each of the following as the main method of contraception in a developing country's family planning programme:
   a) the IUD
   b) sterilisation
   c) the contraceptive pill.

5. A childless couple in their late twenties requesting a sterilisation would be faced with intense questioning from their doctor. Suggest some questions they could be asked to establish their suitability for the operation.

6. Suggest some reasons why people might regret a sterilisation operation and want it reversed.

ACTIVITY

Suitable contraceptive methods

In groups of two or three, discuss the situations described below.

Use the information in this unit to work out which contraceptive methods would be most and least suitable in each case. Give reasons for your choices.

1. Jane (23 years old) and Terry (24 years old) have two children. They agree that they do not want any more.

2. Abigail is 38. She smokes thirty cigarettes a day. She has no children.

3. Mr and Mrs Roberts are in their early forties. They have decided that their family is complete.

4. Peter is recovering from several years of heroin addiction. He has recently got engaged.

5. Laura is 25 years old and a non-smoker with no children. She wants a reliable contraceptive method that does not interfere directly with making love.

6. Nasira has a history of pelvic inflammatory disease and has high blood pressure. She is 40 years old and has one child.

7. Frances is 35. She gave birth to her second child twelve months ago. She says: 'I'm not sure if I want another child. I do not like the idea of interfering with my body by taking artificial chemicals. I've tried the IUD but it gave me terrible cramps and I had to have it removed.'

9 WORLD POPULATION – PROBLEMS AND SOLUTIONS

UNIT 9.4 Population control policy

For couples in poor or rural communities the benefits of rearing large families exceed the costs. The move from an agricultural, rural economy to an industrial, urban economy acts as a good means of population control by pushing up the costs of children. This is what happened in Europe during the eighteenth and nineteenth centuries.

In the late twentieth century it is a different story. Many poor countries cannot develop their economies because of their difficulties with uncontrolled population growth. The governments of these countries can implement a range of strategies to encourage couples to limit their family size. But these strategies will not work unless they change the balance so that the cost of raising a large family is greater than the benefits it brings.

ACTIVITY

Strategies for population control

In an imaginary country in the developing world there has been a sharp increase in the population growth rate over the last 20 years due to improvements in agricultural techniques and health services. The government is trying to decide on the measures to incorporate into its population control programme to reduce the country's birth rate.

Study the possible strategies listed in the box and then work in groups of no more than four people to answer the following questions.

1 Discuss the problems the government is likely to face if it tries to introduce each option.

2 Which ones are likely to be unacceptable to the people?

3 a) Choose the three options that you believe the government should implement, bearing in mind your assessment of the effectiveness, cost and acceptability of each one.
   b) Explain your choices to the rest of the class.

4 If you were put in charge of an advertising campaign in a developing country to promote smaller families, which medium would you use: posters, radio, television, lectures or leaflets? Explain your choice.

5 Devise a script for a commercial or design a poster that could be part of this campaign.

Proposed strategies for a population control programme

1 An advertising campaign to persuade couples that they should have smaller families.

   This could be relatively cheap depending on the media used. No opposition to such a campaign would be expected.

2 Setting up a large number of family planning clinics throughout the country to provide free contraceptive advice and supplies.

   This could be expensive and some opposition may be encountered from religious groups.

3 Launch a 'social marketing' programme to encourage all small village general stores to sell condoms at a very low subsidised price. Back this up with a widespread advertising campaign to encourage people to buy them.

4 Legislate to ban all marriages for men under 24 and women under 21.

   This highly controversial but inexpensive measure has been attempted in China. It does not change attitudes to family size but restricts the timespan for legitimate births.

5 Scrap the system of welfare benefits for families with children and increase taxes on parents who have large families.

   This strategy would increase government revenues but it is likely to have an adverse effect on child welfare. However it has been successful in reducing the birth rate in Singapore.

Science, Technology and Society © D. Andrews, 1992

6 Introduce compulsory secondary schooling for all children up to the age of 16.

This is an expensive measure but it has many other benefits. By indirectly reducing the economic value of children as farm labour, it tackles the motivation for large families at its roots. However, it may cause financial hardship to the families affected.

7 Encourage people to be sterilised by giving away expensive free gifts to all who agree to have the operation.

An incentive scheme like this was tried in India but it aroused considerable opposition as an element of coercion was sometimes employed by over-zealous government officials. If it is handled carefully and there is a high take-up it could be successful.

8 Expand the career opportunities for women to be employed outside the home.

This indirect approach involves changing attitudes and expanding the economy, which may prove difficult.

9 Introducing a state pension to give elderly people a guaranteed income.

This could be expensive, depending on the population's life expectancy. It removes the incentive for viewing children as future providers for parents' old age.

ACTIVITY

Changing family size

In the UK the average family size has changed over recent years and it will continue to change in the future. In this activity you can measure this change and explore the influences behind it.

1 Ask each pupil in the class for the following information and record it in a suitable table.

a) How many children were born to your maternal grandparents?

b) How many children were born to your paternal grandparents?

c) How many children are there in your family?

d) What would be the ideal number of children you yourself would like to have?

2 Display the data so that the differences between the four groups can be readily identified.

3 What is the main difference between the three generations' family sizes?

4 Attitudes to the family have changed and are continuing to change. How do you think the following have influenced changes in family size over successive generations?

a) A desire for improved living conditions.

b) Women's desire to pursue a career outside the home.

5 What other factors do you think have led to the reduction in family size?

10 A PILL FOR EVERY ILL?

UNIT 10.1 Drug development: the sulphonamide story

Bullets and shells were not the only threat to the soldiers fighting World War I. In the appalling conditions of trench warfare many lives were lost because there was no treatment to combat fatal infections such as gangrene and cholera. After the war medical scientists from the UK and Germany, who had direct experience of the horrors of the front line, returned to their laboratories fired with enthusiasm to discover substances that could destroy the bacteria that caused infectious diseases such as pneumonia and tuberculosis.

The story of Alexander Fleming, the Scottish scientist who discovered the antibiotic penicillin in 1928, is well known. Just as important, but less famous, is the story of the German medical researcher, Gerhard Domagk, who discovered **sulphonamide**, the first antibacterial drug produced by chemical synthesis techniques in the laboratory. He began the search in 1929 with the assistance of his colleagues Josef Klarer and Fritz Mietzsch. The following account is an excerpt from an article published in *research*, the scientific magazine of Bayer AG Leverkusen, the chemical company that Domagk worked for.

Winning the Fight Against Bacteria

The scene was Christmas of 1932. After five years of painstaking work with many disappointments, one of the compounds that Josef Klarer had synthesized from azo dyestuffs – the so-called sulphonamide – showed a strong antibacterial reaction. "The best chemotherapeutic effects on a streptococcal infection ever seen in an experiment with animals," as the usually more down-to-earth Domagk enthusiastically concluded. The team working with the untiring researcher kept up an enormous pace during the days between December 20 and 28. "We performed dissections until we couldn't stand up any more. We studied the pictures of our microscopes until we couldn't see. And then suddenly, as if we had been hit by lightning, the significance of the results became clear." While the untreated animals died from the infection within two days, all of those animals which had been treated with sulphonamide survived.

Without the help of vivisection, so-called *in vivo* experiments, the effect of this substance with the chemical name 4-aminobenzene sulphonamide could not have been discovered using the know-how and the methods available in those days. "*In vitro*" experiments, i.e. in a test tube, showed absolutely no effect at all with sulphonamides.

After completing the animal experiments, Domagk released the substance under the designation "Streptozon" (it would later be marketed under the trade name of Prontosil) for clinical testing. The first results went beyond all hopes, but not all tests were successful because the dosage was too low. The therapeutic results proved to be excellent only after Fritz Mietzsch was able to synthesize it in an improved form which made higher dosages possible. When the results were finally published in 1935, the medical world had its sensation. However, there were also sceptics, as an amused Domagk later reported: "When we wanted to introduce the chemotherapy of streptococcal infections to a clinic, a man who would one day be a full professor claimed that he had tried to verify my experiments without success and that the whole sulphonamide fuss would very quickly pass."

The professor proved to be wrong. Within a short time, Prontosil won recognition all over the world. Enthusiastic headlines in the U.S. press praised the new discovery from Germany in 1936: the sulphonamides had reportedly saved the life of U.S. President Franklin D. Roosevelt's son, who had been suffering from a throat infection.

In 1939 Gerhard Domagk was honoured with the Nobel Prize.

Excerpt from *research*, the scientific magazine of Bayer AG, Leverkusen, Germany.

ACTIVITY

Sulphonamide

1 Explain the meaning of:
   a) streptococcus cultures
   b) chemotherapeutic effects
   c) vivisection
   d) therapeutic results.

2 What is the difference between *in vivo* and *in vitro* experiments?

3 What was the main justification for using *in vivo* experiments in the search for the antibacterial substance?

4 As a child Domagk had a reputation for meticulous observation and the ability to register the smallest unusual detail in the world around him. Why were these skills useful to him in his chosen career?

Science, Technology and Society © D. Andrews, 1992

10 A PILL FOR EVERY ILL?

UNIT 10.2 Testing drugs on humans and animals

It can take up to fifteen years and £150 million to develop a substance into an effective drug that is safe for human use. In this lengthy and expensive process the pharmaceutical research scientists are guided by the principle first set down by the Greek physician Hippocrates:

'Two rules apply when treating disease: to help the patient and to cause no injury.'

To follow this principle new drugs are tested carefully to demonstrate their efficacy in tackling the specified illness and to ensure that the risk to the patient from undesirable side effects is minimal. This includes initial testing on animals, a practice that generates much debate, and human clinical trials, before the government will give the drug a licence for use. However this system is not foolproof and sometimes drugs with harmful side effects have to be withdrawn from the market.

Human drug trials

In a clinical drug trial the patients participating are allocated to one of two groups: members of the experimental group are given the drug treatment while those in the control group are given an inactive substance (a placebo). This ensures that any differences between the two groups can be attributed to the test drug's action on the patient. The responses of the two groups are then carefully recorded and compared. Some patients in the control group will recover spontaneously without the need for treatment and some will feel better because they believe that they have been taking a drug to help them recover. This beneficial psychological influence on the person's well-being is known as the 'placebo effect'.

ACTIVITY

Your views on human drug trials

1. Suggest why the following requirements are important in the clinical testing of drugs on humans.
   a) The control and experimental groups should be matched so that they are similar in terms of age, sex and the severity of the disease.
   b) The patients in the trial are unaware of whether they are in the control or experimental group ('single blind' test).
   c) Both the patients and the doctors participating in the trial are unaware whether the patients are in the control or experimental group ('double blind' test).

2. 'Show me a drug without side effects and I'll show you a drug that does nothing' is the view of one doctor. Are you prepared to accept the minor side effects of a drug?

3. In some cases the impact of a drug's side effects can be reduced by restricting the circumstances in which it can be used, or by limiting the patient population to which it can be prescribed. Check the labels of the drugs you have at home. Does this apply to any of them?

4. In the UK, the Committee on the Safety of Medicines (CSM) is responsible for collecting information on the adverse side effects of prescribed drugs. What different items of information would it need to collect about the patient and the side effects to be effective in its role?

5. The CSM's monitoring system depends on doctors voluntarily sending in a yellow report card when one of their patients suffers from serious side effects of a drug that they have taken. What might be the drawbacks of relying on a system like this?

① Arguments in favour of the use of animals in drug testing

- There is a need for new drugs to be developed to tackle the many illnesses that do not at present have an effective cure, such as cancer, heart disease and AIDS.

- It would be too risky to carry out the testing of an unknown drug on humans, who might suffer serious side effects or even die.

- The risk to an animal's health is acceptable when it is balanced against the benefits of discovering a life-saving treatment, or the need to protect a human life from permanent damage due to the unforeseen side effects of a drug.

- If the drug thalidomide had been tested on pregnant animals then its harmful effects on fetuses would have been detected and we could have prevented the unnecessary tragedy of thousands of babies being born severely disabled.

- Whenever possible drug testing is carried out on tissue cultures. Unfortunately the action of the drug when it enters the organism's body can be very different.

- Researchers do not wish to cause deliberate and unnecessary harm to animals. The law requires that the pain experienced by an animal in an experiment is kept to an absolute minimum.

- We must accept that animals sometimes may have to die for the greater benefit of saving human lives.

② Arguments against the use of animals in drug testing

- Drug treatment does very little to save human lives. The reduction in the numbers of deaths from infectious diseases had been achieved before the introduction of drugs and vaccines.

- In developing countries more people's lives could be saved if the money spent on drugs was instead used to provide clean water and an adequate diet.

- The use of animals is of no value in drug testing because all species have their own physiological characteristics and so react differently. The guinea pig, for example, will die immediately if it is given penicillin.

- Many drugs that have been passed as safe after testing on animals have later had to be withdrawn because of serious and sometimes fatal side effects. For example, Opren, the anti-arthritis drug, was withdrawn in 1982 after causing 70 deaths.

- Many diseases like cancer and heart disease are linked to life-style and are therefore preventable. This makes drugs unnecessary.

- The artificial diseases that are induced in laboratory animals are not the same as the diseases suffered by people in real life.

- There are 18 000 licenced drugs on the market in the UK, yet the World Health Organisation says that only 200 of these are essential for health. The reason the drugs industry is flooding the market with unnecessary drugs is to ensure that its profits are maintained.

ACTIVITY

Your views on drug testing on animals

Study carefully the arguments for and against the use of animals in the development of new drugs in Boxes 1 and 2. Which side seems most convincing to you? Explain your view to a partner.

10 A PILL FOR EVERY ILL?

UNIT 10.3 Tackling the drug problem

One class of drugs – **psychotropic drugs** – can affect our brain cells and alter the way we think, feel and behave. They may be prescribed by a doctor to treat mental illness. Anti-depressants and tranquillisers are the most commonly prescribed psychotropic drugs. People also use psychotropic drugs for their pleasurable effects – alcohol and nicotine are those most commonly used. Psychotropic drugs have many uses but they all suffer from one major drawback. If used excessively over a long time they create dependency in the drug user. Table 10.3.1 describes the effects of some psychotropic drugs.

DEPRESSANT AND STIMULANT DRUGS

Alcohol, cannabis, tranquillisers and solvents belong to a class of drugs called **depressants**, which slow down the activity of the nervous system and heart and breathing rates. Large doses may seriously depress the medulla of the brain, which controls breathing, and this can lead to drowsiness and sometimes unconsciousness. Every year about 40 people in the UK die because their rapid consumption of a large quantity of alcohol totally depresses the breathing centre, with fatal consequences.

Stimulants have the opposite effect, speeding up the activity of the heart and brain. The effect of mild stimulants, such as caffeine found in tea, coffee and some fizzy drinks, is particularly noticeable if it is taken late in the evening, making it difficult to fall asleep. More powerful stimulants, such as cocaine and amphetamines, reduce appetite and produce feelings of energy and confidence. In large doses they cause restlessness, anxiety and insomnia and, in some cases, a state of mental confusion is produced, similar to severe paranoid mental illness.

DEPENDENCY ON DRUGS

If a drug is taken regularly, over a period of time the dose will need to be increased to achieve the same effect. For example, the size of a normal dose injected by an experienced heroin user would be sufficient to kill someone who had never taken the drug. This adjustment in the body's physiology is described as **drug tolerance**. Drug dependency, or addiction, is revealed when someone wishes to stop using a particular drug and finds that when they do so they suffer from unpleasant withdrawal symptoms. To avoid the physical withdrawal symptoms the addict must cut down the drug dose very gradually. Drug dependency can also be psychological rather than physical, or a combination of the two. Psychological dependency develops when the drug-taker comes to believe that he or she needs the drug in order to cope with life stresses.

ALCOHOL: A SERIOUS DRUG PROBLEM?

Alcohol is a legal and, to most people, socially acceptable drug in the UK, although in some countries and religions it is banned. In the UK about 90 per cent of adults drink alcohol, most of them at a moderate level that has a proven beneficial effect on the health. But like any drug, alcohol can have harmful effects on the body. It is thought that one in ten adults is suffering health problems caused by drinking too much. Heavy drinking over several years can lead to alcoholism, a state of dependency in which the person's relationships and work are severely affected. An alcoholic who stops drinking suddenly will experience very unpleasant withdrawal symptoms, sometimes including frightening hallucinations.

TRANQUILLISERS: THE LONG-TERM PROBLEMS

We can usually find strategies to cope with the stress that we experience, but if these coping mechanisms fail we can suffer the unpleasant symptoms of extreme anxiety such as dizziness, sweating excessively, irregular heart beats (palpitations) and sleeplessness. In some cases a doctor will prescribe a minor tranquilliser, such as Ativan or Valium, to reduce these symptoms. These drugs belong to a chemical group known as the benzodiazepines, which block the synapses of the brain's neurones so that the flow of nerve impulses is reduced. Tranquillisers provide short-term help for someone in a crisis, but if taken continuously for several years they result in dependency, which creates withdrawal symptoms that are sometimes worse than the original anxiety. In the UK thousands of people who have been taking tranquillisers for over fifteen years now need long-term support to help them to escape from their dependency. The newspaper report highlights the problems of drug dependency in residential homes for elderly people.

Crackdown on Drugs for the Aged

Old people in residential homes are in danger of being drugged with tranquillisers and sleeping pills by unqualified staff who want an easy life. This warning came yesterday from a report by a working party representing general practitioners from across the country. It highlights the problems of the over-prescribing of drugs to elderly people.

One member of the working party also expressed concern that some doctors were providing bulk prescriptions to old people's homes. One doctor had given a prescription for 1000 sleeping pills. These were given to the elderly residents at the discretion of the unqualified person in charge of the home. He said that it was scandalous that someone without medical knowledge could dole out pills according to personal whim.

A spokesman for MIND criticised the widespread prescription of minor tranquillisers. All too often old people's symptoms of confusion were being misdiagnosed as senile dementia. In fact their behaviour was a result of being topped up with drugs. Because of over-prescribing old people were having bad falls and even becoming addicted to the drugs they were given.

ACTIVITY

Drug problems and solutions

1 Which do you think is a more serious drug problem – alcohol or heroin? Explain your answer.

2 What measures would you like to see introduced to achieve a reduction in the level of alcohol-related problems?

3 Carry out a survey on cigarette smoking among young people. Include questions about the amount smoked, the reasons for smoking and the problems experienced in trying to give up. Use your results to discover whether there are significant differences between male and female smoking behaviour. Draw up a series of recommendations that would help young people in your school or college to change their smoking habits.
You may prefer to survey other types of drug-taking, e.g. alcohol.

4 What needs to be done to tackle the problems described in the newspaper article?

Table 10.3.1 The short-term effects and health risks of some psychotropic drugs.

Type of drug	Short-term effects	Health risks of excessive use
Alcohol	Slowed reactions, distorted judgements, drowsiness, clumsiness	Liver infections (hepatitis), cirrhosis of the liver, Vitamin B deficiency, destruction of nerve tissue, memory loss
Nicotine (in tobacco)	Increased heart rate and blood pressure, breathlessness after mild exercise	Cancer of tongue, throat and lungs, heart disease, bronchitis
Cannabis (marijuana, spliff, blow, pot)	Slowed reactions, coordination is less precise, sometimes hallucinations are experienced	Increased risk of lung cancer, lack of motivation and interest in others
Hallucinogenic drugs (e.g. LSD, Ecstasy)	Distorted perceptions of time and space, hallucinations, distorted mental and emotional state	Fatal accidents due to false beliefs in 'superhuman powers' e.g. jumping from buildings. Vivid 'flash-back' hallucinations can occur many months after stopping use of LSD
Glues and solvents	Slowed reactions, uncontrolled behaviour, sometimes unconsciousness and vomiting	Damage to kidneys and liver, suffocation due to vomit in airways
Heroin	Powerful pain-killer, produces pleasurable sensations	Infections from shared hypodermic needles e.g. AIDS, hepatitis. Risk of fatal overdose. Painful withdrawal symptoms occur if regular use stops suddenly

10 A PILL FOR EVERY ILL?

UNIT 10.4 Drugs in the developing world: inappropriate technology?

Preventative health measures such as adequate nutrition, hygienic sanitation methods, a supply of clean drinking water and a comprehensive mass immunisation programme can achieve very effective reductions in the incidence of common illnesses in the Third World. Curative methods are less useful. There is little point in using drugs that ease the symptoms of illness if no attempt is made to tackle the root causes of disease.

One example of this inappropriate use of technology is the prescribing of Western-style drugs to treat the most common diseases in the Third World – intestinal infections. These illnesses result in a dangerous loss of body fluid through chronic diarrhoea. The alternative method of treatment is to replenish lost fluid and salts directly by drinking a mixture of clean water, sugar and salt. This method – **oral rehydration therapy** (ORT) – is more effective and cheaper than using drugs and does not require the help of a medical expert.

Poor growth in children is another major health problem in developing countries. Advertising campaigns have persuaded some parents to buy drugs in a misguided attempt to improve the growth of their children. The newspaper article describes the way an anabolic steroid, Fertabolin, was marketed as a growth-promoting drug, and some of the criticisms this campaign received.

ACTIVITY

Fertabolin

Read the newspaper article and answer the following questions.

1 For what reasons are anabolic steroids commonly used in Western countries?

2 Explain how Fertabolin was marketed in Third World countries.

3 Suggest why people in developing countries are attracted to spending money on Western-style drugs rather than on food.

4 Expenditure of at least £270 million a year would be required to provide clean water supplies for the rural population of India. This is roughly equivalent to the annual investment by the drug companies in India. What does this tell you about the profitability of providing clean water?

5 The use of herbal medicines has a long and respected tradition and it is still a popular approach in developing countries. In Bangladesh, for example, wood apples that grow wild are an essential ingredient of a widely used remedy for the diarrhoeal diseases that are rampant during the monsoon season. Discuss the advantages and disadvantages of using traditional herbal treatments compared to Western drug treatments.

Hungry Children Victims of Drug Pushers

by Anton La Guardia

A company is profiting from the malnutrition of mothers and children in third-world countries by selling a drug that is claimed to boost growth and well-being, when in fact it is worthless for this purpose. The drug may cause the premature growth of sexual hair, enlargement of the penis or clitoris and other side-effects in children.

The drug, Fertabolin, is being promoted in India with a picture of two children and the caption: "Fertabolin. For a life full of fun and frolic... Helps gain normal weight and height." And in the Philippines it is advertised with a picture and a caption saying it is "safe even for children and infants".

A sample of Fertabolin was bought without prescription in India last month by a worker for Oxfam, the famine-relief agency. A leaflet sold with the drug says: "Fertabolin is an ideal preparation for promoting the general well-being of convalescent patients suffering from general weakness, fatigue, listlessness and anaemia." The leaflet also says that the drug can be given to patients recovering from malnourishment, mothers after giving birth and people suffering from "fatigue due to overwork".

Fertabolin contains a hormone, ethyl-oestrenol, which is sometimes taken in Western countries by weight-lifters who want to develop muscle. However, the British National Formulary, an authoritative drug guide for doctors, says the use of this and similar hormones as body builders or as tonics "is quite unjustified". The drug may cause liver cancer, heart attacks and high blood pressure and may also stunt the growth of children by causing the growing ends of bones to close prematurely.

Michael Rawlins, professor of clinical pharmacology at Newcastle University, says: "There is no indication that these drugs make people gain weight faster. What malnourished children need is food, not drugs." He adds: "The long-term effects on children are unknown. Their use of these drugs would be legitimate only under exceptional circumstances."

Oxfam is pressing for tight international controls on drug sales. It is worried about poor people wasting their meagre earnings on useless drugs. The sample of Fertabolin from India cost 16½ rupees, enough to buy food for four days for a poor family of four to five people. Dianna Melrose, an Oxfam worker, says: "The tragedy is that some people have even sold their land to buy totally unnecessary drugs, which makes their food problems even worse."

Fertabolin is not included in the list of 232 drugs considered to be essential for the Third World by the World Health Organisation, and was banned in Bangladesh in 1982. One drug company, the Swiss firm Ciba-Geigy, announced in April 1982 that it would stop production of a similar drug, Dianabol. Ciba-Geigy said: "The inappropriate use of this product constitutes a risk that far outweighs the benefit of the product's legitimate use."

Organon sells Fertabolin (and other "anabolic steroids") to 29 countries and controls about a third of the market for these drugs, worth some £35m. The company says anabolic steroids are useful and relatively safe.

In December 1983, the Dutch pharmaceutical association, Nefarma, ruled that in the past Organon had put patients, including children, at risk through careless marketing of anabolic steroids. Nefarma said that companies selling drugs in the Third World, where controls are weaker, have an added responsibility.

Organon told Nefarma it had changed its marketing methods. The company says: "Organon cannot be held responsible for misuse which may occur when a doctor's prescription is evaded, nor for the lack of patient information in those cases."

Organon says its advertisements are aimed exclusively at doctors who also have access to consistent information through the company's "product safeguard" a statement given to doctors worldwide. Organon also says it is updating leaflets for all its drugs sold in India. But it could not say what changes would be made to the Fertabolin leaflet.

Oxfam says that while companies profit by selling inessential vitamins, tonics and cough syrups, essential drugs for the Third World are in short supply.

© *The Observer*

11 ENERGY – RESOURCES AND CONSUMPTION

UNIT 11.1 Energy basics

It's a basic truth that nothing can happen without energy. As things happen energy is converted from one form to another (see Figure 11.1.1). During an energy conversion the amount of energy input is the same as the energy output. The First Law of Thermodynamics states that energy cannot be destroyed or created. But if energy is never destroyed how can we say that the world is suffering from an energy shortage?

To answer this we need to distinguish high-grade energy from low-grade energy. It is easiest to think of low-grade energy as matter with a relatively low temperature. These two types of energy are measured in the same units, kilojoules. However, 100 kilojoules (kJ) of low-grade energy is of little use to us compared to 100 kJ of high-grade energy. High-grade energy, such as the chemical energy in fossil fuels, has the potential for producing useful power, whereas low-grade energy cannot be used in this way. A gas burner (high-grade energy) will bring a pan of water to the boil but the boiled water (low-grade energy) cannot be used to heat more water to boiling point. In every energy conversion some high-grade energy is converted into low-grade energy as heat. As a result the total amount of low-grade energy in the universe is increasing. This is the Second Law of Thermodynamics.

These ideas have important implications for the way we use our energy resources. It is important to maximise the efficiency of energy conversions. Efficiency is defined as the amount of useful energy produced as a proportion of the total energy input. A modern coal-fired power station has an efficiency of about 35 per cent. This means that 35 per cent of the chemical energy in the coal is converted into electrical energy. The rest is lost as waste heat (see Figure 11.1.2). Minimising heat losses is one of the best ways of increasing the efficient use of energy. Overall efficiency of energy use can also be improved by cutting down on the number of conversions that take place when a fuel is burnt.

ACTIVITY

Energy flow diagrams

1 Using the names for the different energy forms given in the box, draw an energy flow diagram like those in Figure 11.1.1 for each of the following conversions:
   a) generating electricity at a hydroelectric power station
   b) a battery powered cassette player
   c) a solar powered calculator.

2 a) Draw energy flow diagrams for these energy conversions, which both use gas to provide domestic hot water:
   - Gas used directly to fuel a home's boiler.
   - Gas used as fuel in a gas turbine power station to generate electricity to power a home's water immersion heater.
   b) Which method makes the more efficient use of natural gas?

A car screeches to halt when the driver brakes

Kinetic energy → Heat
            ↘ Wave energy (sound)

An electric door bell is operated

Electrical energy → Kinetic energy → Wave energy (sound)

A steam engine

Chemical energy → Heat → Kinetic energy

Figure 11.1.1   Energy flows.

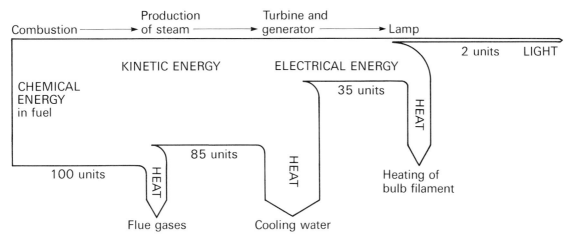

Figure 11.1.2 Energy transfer in the combustion of fuel in a power station to generate electricity that is converted to light and heat in a bulb.

Energy forms

Energy forms are of two types. Potential energy forms store energy that can be released at a specified time. A stretched rubber band is an example of mechanical potential energy.

Kinetic energy is released as things move. For example, electricity can be seen as the flow of electrons.

Potential energies

Chemical bond energy
Gravitational energy
Magnetism
Nuclear energy
Mechanical potential energy
Thermal energy (heat)

Kinetic energies

Kinetic energy of moving objects
Wave energy (sound, light, radio waves)
Electrical energy
Thermal energy (heat)

11 ENERGY – RESOURCES AND CONSUMPTION

UNIT 11.2 Fossil fuels

The fossil fuels, oil, coal and natural gas, have provided the impetus for much of the advance in technology over the last 200 years. Coal was the fuel that powered the steam engines of the industrial revolution. Oil made possible the internal combustion engine, which launched the era of private motor transport. Gas fuel led to the advances in domestic heating and cooking that have contributed to a leap in living standards. Yet the depletion of fossil fuels could pose a threat to further advance. The shortage of fossil fuels, particularly oil, will begin to have an impact within the next 40 years. Short-term oil supply problems have already resulted from disputes in the oil-producing countries of the Middle East. The amount of fossil fuel in the Earth's crust that remains to be exploited is still uncertain. But there is no doubt that it will run out at some time in the future. Fossil fuels are in finite supply.

Estimating the length of time before a particular fossil fuel runs out is very difficult. It depends on an accurate assessment of two independent factors: the amount of the fossil fuel reserve that is yet to be extracted and the future pattern of consumption.

CATEGORIES OF FOSSIL FUEL RESERVES

Some reserves of fossil fuel in the Earth's crust are more easily extracted (producible) than others. For example some may be nearer the surface. At current fuel prices it may be uneconomic to attempt to extract some of the reserves. However this may change if fuel prices increase. To take account of this, fossil fuel reserves are officially classified in three categories:

Proven: reserves which on the available evidence are virtually certain to be technically and economically producible (90 per cent chance).

Probable: reserves which are not yet proven but which have a better than 50 per cent chance of being economically and technically producible.

Possible: not yet regarded as probable but estimated to have a 10 per cent chance of being technically and economically producible.

Table 11.2.1   Reserves of fossil fuels on the UK continental shelf in millions of tonnes.

	Proven	Probable	Possible
Oil	1760	600	640
Gas	1266	654	652

ACTIVITY

Reserves of fossil fuels

Study the two quotes in the box and compare them with the graphs in Figures 11.2.1 and 11.2.2, which show possible patterns of future consumption for oil and coal.

1 According to the graphs, how many years' supply of
   a) coal
   b) oil
   remain in the Earth at present?

2 Give two reasons why the predictions in *A Blueprint for Survival* were inaccurate.

3 In what respect is Odell's statement contradicted by the information in the graph?

4 If the price of fuel increases dramatically, what is likely to happen to
   a) the rate of fuel consumption?
   b) the ratio of proven to probable reserves?

5 What is likely to happen to the price of oil after 2030? Give your reasons.

6 What problems would reduce the chances of a reserve being technically producible?

ENVIRONMENTAL IMPACTS OF FOSSIL FUELS

Fossil fuel use has a considerable impact on air pollution. Combustion of any type of fossil fuel releases carbon dioxide into the atmosphere and adds to the greenhouse effect (see Unit 11.9). Natural gas is the cleanest fuel as it produces the least CO_2 (about 75 per cent less than the heat equivalent mass of coal) and only one other waste gas – water vapour.

In contrast, coal produces a wide range of other toxic gases including sulphur dioxide, heavy metals such as lead and mercury, and radioactive elements such as uranium. Domestic coal fires can produce high concentrations of sulphur dioxide in cities. This caused the deaths of 4000 people from bronchitis in five days during the London smogs of December 1952. Today sulphur dioxide contributes to the formation of acid rain, which has killed off the fish in hundreds of Scandinavian lakes and caused damage to large areas of European coniferous forest.

Burning oil gives off carbon monoxide and the carcinogenic compound benzene. It also releases nitrogen oxides that add to the problem of acid rain and ozone, which directly damages plant life. Lead is also a dangerous component of petrol exhaust fumes (but the use of unleaded petrol is reducing atmospheric lead, see Unit 11.7).

It is not only combustion of fossil fuels that has a harmful impact. Environmental damage also results from their extraction from the Earth's crust. For example, coal mining creates unsightly slag heaps and polluted waterways. After the coal has been extracted, ground subsidence may occur, damaging buildings and underground pipes.

> 'If these rates continue to grow exponentially, as they have done since 1960, then natural gas will be exhausted within fourteen years and petroleum within twenty years.'
>
> from *A Blueprint for Survival* (Penguin, 1972)
>
> '... the oil resource base ... gives very little cause for concern, not only for the remainder of this century but also thereafter well into the twenty-first century at rates of consumption which will then be five or more times their present level.'
>
> Professor Peter Odell (1972)

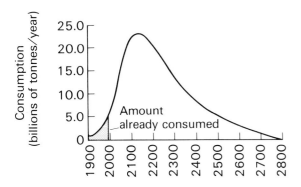

Figure 11.2.1 The theoretical future consumption of coal.

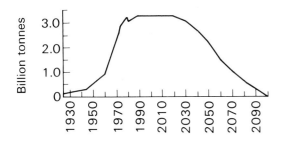

Figure 11.2.2 The theoretical future consumption of oil. It assumes that a plateau of 3.3 billion tonnes per year will be reached in the 1990s.

ACTIVITY

Environmental impacts of fossil fuels

1. Suggest why a switch from coal-fired to gas-fired power stations has been suggested.

2. Describe the environmental risks and problems associated with the extraction and transport of oil, gas and coal.

3. Find out what steps can be taken to remove pollutants from the waste gases of
   a) coal-fired power stations
   b) petrol and diesel engines.

11 ENERGY – RESOURCES AND CONSUMPTION

UNIT 11.3 Supplying electricity

Until 1989 the Central Electricity Generating Board (CEGB) was the state-owned organisation responsible for the production and supply of all electricity in England and Wales. In 1989 the CEGB was split into three separate companies: PowerGen and National Power took over the non-nuclear power stations and were privatised; Nuclear Electric took over the nuclear power stations and remained in public ownership. The data given in this unit is for the electricity produced by the various types of power station owned by the CEGB prior to privatisation.

ACTIVITY

Electricity supplies

Study the data in Table 11.3.1, which gives a summary of the CEGB electricity generation in the financial year 1987–88.

1 Calculate the contribution of each of the following types of power station to the total amount of electricity supplied by the CEGB and display your data in suitable graphical form:
   a) nuclear power stations
   b) coal-fired power stations
   c) oil-fired power stations
   d) dual (oil/coal) power stations
   e) gas-fired stations
   f) hydroelectric: natural flow.

2 Why is it useful to know a power station's load factor? Suggest why it never reaches 100 per cent.

3 Which type of power station has
   a) the highest load factor
   b) the lowest load factor?

4 Suggest reasons for this wide variation in load factor.

5 Use evidence from Table 11.3.1 to show that larger power stations are more efficient than smaller ones.

6 What percentage of the energy value of fuel consumed in all types of power station is not converted into electricity? What has happened to this energy?

7 Why is there no figure for the thermal efficiency of hydroelectric power stations?

8 The contribution of hydroelectric power in Scotland is about 10 per cent of the total. Explain why this is different from the percentage for England and Wales.

9 Which type of nuclear power station has the highest thermal efficiency? Suggest a possible explanation for this (see Unit 13.2).

Science, Technology and Society © D. Andrews, 1992

Table 11.3.1 Summary of electricity generation in England and Wales by CEGB power stations (1987–1988).

Type of station	Declared net capability[1] (MW)	Electricity supplied (GW h)	Load factor[2] %	Thermal efficiency[3] %
NUCLEAR PLANT				
Magnox	3 499	22 664	73.7	25.40
Advanced gas-cooled	1 570	10 208	35.7	34.78
FOSSIL-FUELLED PLANT				
Coal-fired				
500–660 MW stations	20 228	126 426	71.0	36.54
less than 500 MW	11 024	45 066	46.4	32.94
Dual-fired (oil/coal)				
500–600 MW	1 920	9 191	54.5	34.69
less than 500 MW	2 584	6 919	30.5	32.10
Oil-fired	8 417	8 034	12.5	34.48
Gas-fired	2 517	49	0.2	19.34
HYDROELECTRIC STATIONS				
Natural flow	107	249	25.3	–
Pumped storage	2 088	(659)[4]	–	–
TOTAL	53 954	228 147	48.6	33.96

Notes

1 The Declared Net Capability (DNC) is the maximum electrical power generation (in megawatts) that can be maintained indefinitely without damaging the plant.

2 The load factor is the amount of electricity actually supplied during the year as a percentage of the DNC.

$$\text{Load factor} = \frac{\text{Annual electricity supplied (GW h)} \times 100\%}{\text{Hours in the year } (365 \times 24) \times \text{DNC}}$$

3 The thermal efficiency is the percentage of the energy in the fuel consumed in the power station that is converted into electrical energy.

4 Pumped storage stations can consume more electricity than they generate. This figure is a net deduction from the amount of energy supplied.

11 ENERGY – RESOURCES AND CONSUMPTION

UNIT 11.4 International comparisons of energy consumption

As technology has advanced it has involved a greater consumption of energy. Motor transport, heating, lighting and modern household appliances are some of the main ways in which modern Western society guzzles energy. Developing countries do not have such high energy demands at present (see Table 11.4.1). However, as their economies expand their energy demands are likely to increase.

ACTIVITY

Energy consumption

Study Table 11.4.1 carefully before answering these questions.

1. Which four countries are together responsible for almost two-thirds of the world's production of nuclear energy?

2. Compare oil consumption in the USA and the UK. How do you explain the difference?

3. Which is the only country in the table where over half the energy consumed is from coal? Suggest possible reasons for this.

4. Which two countries rely most heavily on natural gas for their energy needs?

5. Compare the patterns of energy consumption in France and the UK, giving reasons for the differences.

6. In which country is hydroelectric power the most significant form of energy? Suggest why this is the case.

7. a) Find out the population sizes of the various countries and regions in the table and calculate the annual energy consumption per head for each country.
   b) Suggest explanations for the differences you find.
   c) Are the differences likely to get bigger or smaller in the future?

8. Which forms of energy generation are not mentioned in the table? In which countries are these other energy-producing technologies likely to make a significant contribution in the future?

Table 11.4.1 Energy consumption in millions of tonnes of oil equivalent for selected countries and regions of the world. (1 tonne of oil ≡ 1.5 tonnes of coal ≡ 400 therms ≡ 1.11 cubic metres of natural gas)

	Oil	Gas	Coal	Hydroelectric	Nuclear energy	Total
USA	724.1	444.5	444.3	82.9	104.6	1799.4
UK	77.8	47.9	61.9	1.3	13.0	201.9
France	83.9	23.3	24.1	12.9	45.1	189.9
Norway	8.8	—	0.6	22.8	—	32.2
Australia	27.0	12.5	35.2	4.0	—	78.7
Africa	82.8	25.9	67.8	17.6	1.0	195.1
Japan	201.3	36.0	72.6	21.8	33.6	365.3
South Asia	53.4	17.5	105.0	19.1	1.1	196.1
SE Asia	112.8	17.6	38.2	6.3	11.4	186.3
USSR	447.7	477.2	361.9	53.5	36.0	1376.3
China	124.3	92.9	330.7	21.5	11.1	580.5
Latin America	209.5	69.9	25.0	75.2	1.3	380.9
TOTAL WORLD	2809.4	1491.8	2277.8	498.2	337.1	7414.3

11 ENERGY – RESOURCES AND CONSUMPTION

UNIT 11.5 Patterns of energy use in the UK

The different energy sources – coal, oil, gas and electricity – have their particular values for specific uses. The domestic, industrial, agricultural and transport sectors have very different energy needs and the nation's energy policy must take this into account. For example, it would be impossible for much of the transport sector to survive if oil supplies were cut off.

ACTIVITY

Study Table 11.5.1, which shows the fuel consumption patterns in the UK, before answering these questions.

1. What percentage of total UK energy consumption is used by
   a) the domestic sector
   b) industry (including iron and steel)
   c) road transport?

2. Which sector uses solid fuel for over half of its requirements? Why does this fuel particularly meet this sector's needs?

3. In which sectors is at least 20 per cent of energy consumed provided by electricity?

4. Which sector accounts for over half the UK's gas consumption? Suggest a possible reason for this.

5. Design a suitable way of displaying this data in graphical form.

6. What changes do you think there have been in the pattern of energy consumption in the different sectors in the last ten years?

7. What changes do you think there will be in the pattern of energy consumption in the domestic sector over the next ten years?

8. Suggest why nuclear power or renewable energy sources do not appear in this table.

Table 11.5.1 UK fuel consumption (10^8 therms), 1986.

Sector	Coal	Gas	Oil	Electricity	Total
Iron and Steel	17.8	7.6	2.5	2.7	30.6
Other industry	21.4	54.2	33.9	26.0	135.5
Domestic	25.9	105.0	9.8	31.8	172.5
Public services	3.8	13.3	10.8	6.3	34.2
Road transport	—	—	135.2	—	135.2
Water transport	—	—	4.4	—	4.4
Air transport	—	—	25.7	—	25.7
Rail transport	—	—	3.0	1.1	4.1
Agriculture	0.1	0.3	3.7	1.4	5.5
Miscellaneous	1.5	16.4	5.4	16.0	39.3
TOTAL	70.5	196.8	234.4	85.3	587.0

Science, Technology and Society © D. Andrews, 1992

11 ENERGY – RESOURCES AND CONSUMPTION

UNIT 11.6 Trends in energy consumption in the UK

Nuclear generated electricity, hydro-electricity and the three different fossil fuels, oil, coal and gas, are described as primary fuels. The energy used in transport, homes and industry can be traced back directly or indirectly to one of these primary fuels.

The graph in Figure 11.6.1 shows that over the last 40 years there have been dramatic changes in the pattern of consumption of primary fuels. In this unit you will be investigating these changes and attempting to explain them.

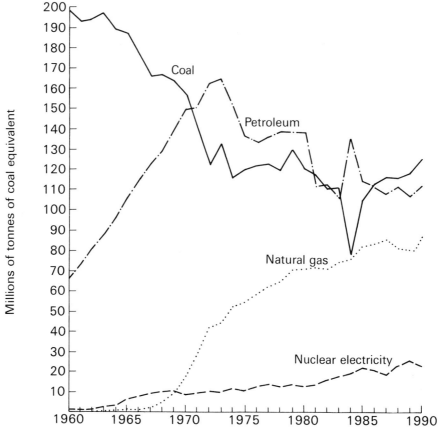

Figure 11.6.1   The UK consumption of primary fuels (1950–1990).

ACTIVITY

Energy trends (1960–1990)

Use the information in the graph in Figure 11.6.1 to help you to answer these questions.

1. Which two primary fuels were not consumed in significant quantities in the early 1960s? Why was the consumption of these fuels so low?

2. How do you think car ownership changed during the 1960s? What effect did this have on the pattern of fuel consumption?

3. Explain how the trend in coal consumption during the 1960s might have been influenced by changes in each of the following:
   a) domestic heating systems
   b) rail transport
   c) the replacement of coal gas by natural gas.

4. a) Calculate the decrease in the consumption of petroleum between 1973 and 1975.
   b) Express this figure as a percentage of the 1973 consumption.
   c) What event in 1974 caused this dramatic reduction in consumption?
   d) In 1980 a revolution in Iran led to a crisis in oil supply and a trebling in oil prices. What impact did this have on oil consumption?

5. There were lengthy nationwide strikes by coalminers in 1974 and 1984. From the evidence of this graph, which of these strikes had the greatest effect on coal consumption?

6. What was the impact of the coal strike on oil consumption:
   a) in 1974
   b) in 1984?

7. The 1974 miners' strike led to electricity power cuts because of the lack of fuel for coal-fired power stations. A similar strike in 1984 did not cause power cuts.
   a) What evidence from the graph explains the different effects of the two strikes?
   b) What political decision had been taken by the government over coal-fired power stations after the 1974 miners' strike?

8. a) What appear to be the major trends in fuel consumption over the last ten years?
   b) Do you think these trends will continue in the future?

ACTIVITY

Energy trends (1977–1987)

Use the data in Table 11.6.1 and Figures 11.6.2 and 11.6.3 to answer these questions.

1. Describe the different trends in the overall energy consumption of the various sectors.

2. Which sector had the highest increase in energy consumption over the ten-year period?

3. How would you explain the differences in the trends seen in the domestic and the industrial sectors?

4. Suggest two possible reasons for the trend seen in the UK's iron and steel industry.

5. a) Describe the major trends shown in Figures 11.6.2 and 11.6.3.
   b) Describe any links between them.
   c) Is there a relationship between the cost of an energy source and its consumption?

Table 11.6.1 Energy consumption (millions of therms) of different sectors in 1977 and 1987.

Sector	1977	1987
Domestic	15 045	17 253
Iron and Steel	4 901	3 052
Other industry	17 914	13 557
Transport	13 051	16 940

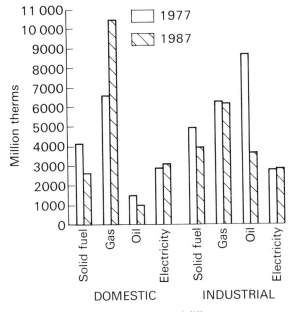

Figure 11.6.2 The consumption of different energy sources by domestic and industrial users in the UK (1977, 1987).

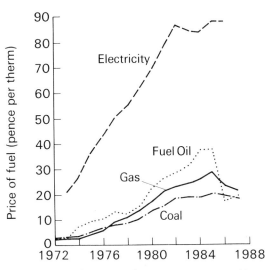

Figure 11.6.3 The costs of energy sources used in manufacturing industry (1972–1987).

Science, Technology and Society © D. Andrews, 1992

11 ENERGY – RESOURCES AND CONSUMPTION

UNIT 11.7 Energy, transport and pollution

All forms of transport use energy. Energy-efficient transport is therefore essential if we are hoping to conserve scarce fossil fuels. Using less fuel in transport will also reduce the pollution that many forms of transport produce in large quantities.

Table 11.7.1 Transport energy trends – energy consumed in millions of therms.

		Coal	Petroleum	Electricity
1977	Road	—	10 176	—
	Rail	17	377	100
	Water	3	521	—
	Air	—	1 857	—
1987	Road	—	13 522	—
	Rail	—	402	105
	Water	—	438	—
	Air	—	2 572	—

LEADED PETROL

Each year human activity releases 450 000 tonnes of lead into the atmosphere, about half of which comes from burning leaded petrol. For many years the oil industry added lead to petrol to help engines run more smoothly. During the 1970s there was increasing concern that lead pollution was harming children's health. The evidence showed that children living near to urban motorways had a high concentration of lead in their bodies. At this level, lead could slow down children's brain development.

During the 1980s a campaign was launched to get the lead removed from petrol. Technological advances in car engine design meant that lead in petrol was not as necessary, and in fact the amount of lead in ordinary petrol had been considerably reduced. Eventually the campaign was successful and in 1987 unleaded petrol became available for the first time in the UK. At first the sales of unleaded petrol were very low. In the March 1989 Budget the tax on unleaded petrol was reduced so that it cost 2 pence per litre less than leaded petrol. Although most new cars can run on unleaded petrol, many older models cannot do so until their engines are adjusted.

ACTIVITY

Transport energy trends (1977–1987)

Table 11.7.1 shows how the energy consumed by the different UK transport sectors changed in the ten years between 1977 and 1987. Study the data it contains to answer these questions.

1. Calculate the increase in the energy consumption by
   a) road transport
   b) air transport.

2. Which of these two has shown the higher percentage increase?

3. Describe the various changes in the pattern of energy consumption by rail transport. Why have these changes occurred?

4. Discuss whether you think the trends in the data are likely to continue into the future.

5. The overall consumption of petroleum by all consumers was 23 440 million therms in 1987. What percentage of this was used up by transport?

6. What major restrictions on the use of private cars would be acceptable in an attempt to cut the emissions of CO_2, nitrous oxides (NO_x) and ozone?

Table 11.7.1 Monthly sales of unleaded and leaded petrol (1988–1991) in millions of tonnes.

	Unleaded	Leaded
1988		
2nd quarter	33	5769
3rd quarter	70	5940
4th quarter	144	5778
1989		
1st quarter	374	5318
2nd quarter	1106	4975
3rd quarter	1481	4634
4th quarter	1687	4349
1990		
1st quarter	1792	4125
2nd quarter	2032	4135
3rd quarter	2178	4033
4th quarter	2253	3762
1991		
1st quarter	2189	3480

ACTIVITY

The switch to unleaded petrol

1 Plot the data in Table 11.7.1 on a suitable graph. Make the necessary calculations to allow you to plot the total sales of petrol for each quarter.

2 What percentage of total petrol sales consisted of unleaded petrol in the first quarter of
   a) 1989
   b) 1990
   c) 1991?

3 Suggest possible explanations for the trends in the sales of leaded and unleaded petrol over this period.

4 Which periods of the year have the lowest total petrol sales? Suggest possible reasons for this.

5 Assume that the sales of unleaded petrol continue to increase at the same rate as they did during the period 1989–1991. Predict when unleaded petrol will:
   a) overtake leaded petrol sales
   b) make up 100 per cent of total petrol sales.

6 Do you think that these predictions are likely to be fulfilled in reality? Explain your answer.

7 Carry out a survey of the petrol prices in your local area to find out the range of price differentials between leaded and unleaded petrol.

8 Do you agree that taxation changes should be used to widen the gap between the prices of these two types of petrol?

9 Apart from tax changes, what other methods could be used to encourage the switch from leaded to unleaded petrol?

11 ENERGY – RESOURCES AND CONSUMPTION

UNIT 11.8 Save it!

Can you reduce your electricity bill? Electricity is an energy source used in practically every home. But do you use it wisely? There are several ways in which you can reduce the amount of electricity you use. You can gain personal experience of the problems by implementing an energy-saving campaign.

ACTIVITY

Checking your electricity consumption

Each member of the class should take readings of the electricity meters in their homes at the beginning and end of an exactly timed period of one week.

1. Why is it important that all members of the group take their readings at the same time?

2. What differences in electricity consumption exist between different homes in your group?

3. How can you explain these differences? For example, is there a link between the number of rooms in a house and the amount of electricity consumed?

ACTIVITY

Carrying out an electricity-saving survey

1. For each room in your home, list the electrical appliances that are used at least once a week.

2. For each appliance, list all the possible ways in which it can be used more efficiently. Include suggestions that may not be easy to put into practice, such as watching less television, or wearing extra clothes instead of using an electric heater.

3. Draw up a list of house rules that would reduce the use of your home's lightbulbs.

4. Investigate other ways in which you could save electricity, such as the use of energy-saving light bulbs (Figure 11.8.1). Friends of the Earth and the Energy Efficiency Office produce helpful leaflets.

ACTIVITY

Launching your Save it! campaign

1. Study your energy-saving ideas carefully. Try to convert these ideas into rules or guidelines that can be understood easily. For example, 'watching less television' could become 'TV-viewing should be restricted to between 6.00 p.m. and 10.30 p.m.'

2. Give each of your guidelines a score from 1 to 5 according to how easy it will be for your family to accept it and act on it. (Very difficult = 5; very easy = 1)

3. Using these scores as a guide, decide whether you wish to modify any of your ideas to make them more acceptable to other members of your family. But remember that you want to save as much electricity as possible.

4. Work out a strategy for ensuring that the rest of your family carry out your SAVE IT! campaign. You will need to consider these points:
   - How can you persuade everyone that these energy-saving measures are worth while?
   - What triggers could you use to remind your family of the energy-saving actions they should be taking?

5. As a class, agree another weekly period during which you will all implement your SAVE IT! campaigns in your homes. Once again, take electricity meter readings at the beginning and end of the week.

ACTIVITY

Following up your campaign

Was your campaign a success?

1. Compare the amount of electricity consumed in your home during the two survey weeks. How much did you save in the second week:
   a) in the number of units of electricity consumed?
   b) in the cost of the electricity consumed?

2. If your family continued with your campaign, what annual saving would they make?

3. Express the amount of electricity you saved as a percentage of the electricity consumption in the first week. Compare your percentage with others in your class. Discuss why some people achieved higher percentage savings than others.

4. Did some people have more success in gaining the enthusiasm and co-operation of other members of their families? What was the key to their success?

Energy saving tips

- Use lagging to insulate water tanks.
- Turn down the hot water tank thermostat – the water should be no more than hand hot.
- Use electric immersion heaters sparingly. It is more energy efficient to use gas.
- Keep the fridge or freezer well-stocked.
- Use cooler washes and full loads in a washing machine.
- Avoid over-use of tumble-dryers. Drying clothes naturally costs nothing.
- When buying new electric appliances, such as televisions and fridges, check that they are the most energy efficient available.
- Do all your ironing in one session, rather than doing small amounts frequently.

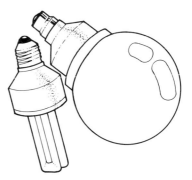

Figure 11.8.1   Types of energy-saving lightbulb.

11 ENERGY – RESOURCES AND CONSUMPTION

UNIT 11.9 Global warming

Governments around the world now recognise that global warming is a serious threat. This unit explores the causes of global warming and the measures that are needed to counteract it.

WHAT CAUSES GLOBAL WARMING?

The 'greenhouse gases' in the Earth's atmosphere form a layer that has the same effect as the glass in a greenhouse (see Figure 11.9.1). It allows the Sun's rays to pass through to heat up the Earth's surface. The hot Earth then radiates heat as infrared radiation at a different wavelength from the incoming rays. This radiated heat is absorbed by the greenhouse gases in the upper atmosphere and partly reflected back to Earth. Without this greenhouse layer the Earth would not be warm enough to sustain life. However, since the industrial revolution of the late eighteenth century, the concentrations of the greenhouse gases have been increasing and the atmosphere is beginning to heat up (Figure 11.9.2).

THE GREENHOUSE GASES

Carbon dioxide is the main greenhouse gas, accounting for about 50 per cent of the global temperature rise. This has occurred because fossil fuel burning (which produces CO_2) has increased and green plants (which remove CO_2 from the atmosphere) have been steadily destroyed. In particular the tropical rainforests have been cut back to provide timber, wood for fuel and grazing land. Open-cast mining for minerals is also taking its toll of the rainforests in parts of the Caribbean and South America (see Unit 2.5).

Methane is produced by the increasing areas of land used as paddy fields. The numbers of cattle have increased enormously to satisfy the demand for beef (particularly in the Western world) and these cattle also produce methane as a result of bacterial action in their digestive systems.

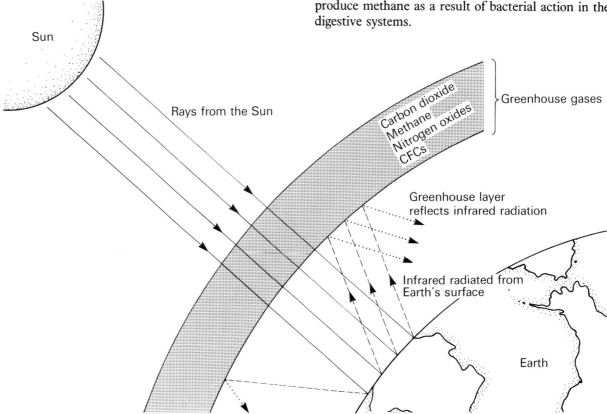

Figure 11.9.1   The greenhouse effect is the result of the build up of certain gases in the upper atmosphere.

Car exhaust fumes contain a significant proportion of nitrous oxides and the motor traffic boom in recent decades has also contributed to the greenhouse effect. The CFC increase can be attributed to its use in aerosols until recent years, and refrigeration systems.

THE IMPACT OF GLOBAL WARMING

In the 30 years between 1960 and 1990 the global mean temperature rose by 0.5°C (see Figure 11.9.2). If the present trend continues it will increase by 1°C by 2020 and 3°C before 2100 and lead to a significant rise in sea level. The melting of the polar ice caps would contribute only a small amount to the rising sea level. The main contribution would be the overall expansion in the volume of water in the oceans caused by the increased temperature. Predicted climatic changes include widespread droughts in Africa and a Mediterranean-type climate in the UK. Unless precautions are taken, low-lying areas could be permanently flooded. Poorer countries, unable to finance the construction of new sea defences, could be particularly hard hit.

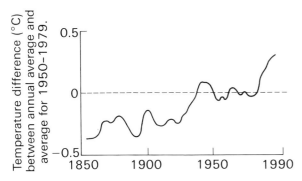

Figure 11.9.2   The rise in the mean global temperature (1850–1990).

ACTION TO STOP GLOBAL WARMING

In 1990 several countries made independent commitments to curb the growth in emissions of one of the greenhouse gases, CO_2. The UK government agreed to a target of stabilising CO_2 emissions at 1990 levels by 2005. Despite international agreement that the problem exists, there are two major questions that remain unanswered:

- Will these reductions be sufficient to prevent the predicted catastrophes of global warming?
- How will these reductions in CO_2 be achieved?

Scientists are divided over the first issue. The second is complicated by a host of technological, political and economic factors. The main measures that have been suggested for tackling global warming are listed right.

Measures to combat global warming

1. Produce electricity using renewable energy technologies instead of fossil fuels.
2. Switch from fossil fuel power stations to nuclear power stations.
3. Improve the efficiency of fossil fuel power stations.
4. Channel waste heat from power stations to heat nearby homes (combined heat and power).
5. Provide economic aid to Third World countries so that they can plant more trees to restore the rainforests.
6. Reduce the amount of road traffic by:
    a) doubling the price of petrol
    b) stopping the construction of new roads and motorways
    c) levying a toll on car drivers entering cities
    d) improving public transport to provide a service that is cheaper, more frequent and more accessible
    e) removing the tax subsidies on cars to stop them being used as a tax-free perk.
7. Improve the energy efficiency of private motor cars by:
    a) reducing speed limits on motorways to 55 mph (90 km/h)
    b) introducing a tax on cars with a high petrol consumption.
8. Improve the energy efficiency of domestic electrical appliances such as fridges.
9. Provide subsidies to encourage industry to improve fuel efficiency.

ACTIVITY

Decisions on global warming

1. In groups, discuss the measures suggested for tackling global warming.
2. List the advantages and disadvantages of each one.
3. Choose the four that you believe should have the highest priority for implementation. Explain and defend your choice to the rest of your class.
4. What other measures can you suggest to counteract global warming?

12 RENEWABLE ENERGY RESOURCES

UNIT 12.1 Solar energy

The solar energy that falls on 50 000 square kilometres of the Earth's surface is more than enough to cater for the world's entire energy needs. This may seem an enormous area but it represents a mere 0.25 per cent of the total area covered by the world's deserts. The problem facing us is to develop the technology to harness this energy.

The Sun's energy can be used directly in several ways.

- **Passive solar design** – the design of buildings to collect, store and distribute solar energy in the form of heat and light. This is regarded as the most effective way of harnessing the relatively low levels of solar energy we experience in the UK. The conservatory containing large panes of glass built on to the rear of a house is an example of such a design.

- **Active solar heating** – making maximum use of the Sun's energy for water heating. In the sunniest areas of the world, solar energy can be concentrated, using large parabolic mirrors, so that it brings water to boiling point. More commonly solar collectors are used to heat water to lower temperatures. Figure 12.1.1 shows how this technology can be used to help heat water for a swimming pool.

- **Photovoltaic panels** – the direct conversion of solar energy to electrical energy. They are only efficient in high light intensities and are therefore mainly used in tropical countries to power a range of equipment from refrigerators to irrigation pumps.

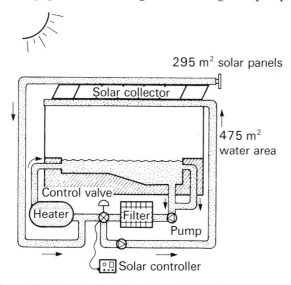

Figure 12.1.1 In the UK, active solar heating systems are particularly well-suited to heating water in swimming pools.

ACTIVITY

Passive solar design

Read the information on passive solar design before answering these questions.

1. Study the housing estate layouts (a) and (b) in Figure 12.1.2. Use the information in the box to decide which one has been designed to maximise passive solar heating.

2. Figure 12.1.3 shows the path of light entering a clerestory in the roof of a building. How is it an improvement on a flat glass skylight?

3. What do you think is meant by 'solar gain'?

4. Explain how the following can influence the solar gain of a house:
   a) a conservatory
   b) large windows on the south-facing side
   c) small windows on the north-facing side.

5. Study Figure 12.1.4 and compare
   a) days 1 and 2
   b) days 2 and 3.
   Suggest explanations for the differences you detect.

6. a) Suggest the heat sources in the home that contribute to the incidental gains shown in Figure 12.1.5.
   b) Why are these heat sources relatively constant throughout the year?

7. From Figure 12.1.5, during which months is solar gain providing over half of the house's heating requirements?

8. From Figure 12.1.5, over the year what percentage of the total heating needs is provided by:
   a) solar gains
   b) auxiliary heating?

9. What do the graphs in Figures 12.1.4 and 12.1.5 tell you about:
   a) the advantages
   b) the disadvantages of passive solar design?

SOLAR ENERGY TECHNOLOGY – PASSIVE SOLAR DESIGN

Any opening in a wall admits sunlight, but the balance between heat gain and heat loss through it depends on its orientation, how it is glazed, how big it is, what the climate is and how the building is heated. Passive solar design (PSD) is the art of manipulating these factors to produce the best possible environment in a building without employing additional energy sources.

Direct Gain

Economic improvements in the energy performance of modern buildings can be achieved by arranging windows and skylights to allow solar energy into areas that require heating. The 'direct gain' passive solar approach typically entails buildings having large windows on the south side and smaller glazed areas facing north. Such features can be incorporated at little or no extra cost.

Sunspaces and Conservatories

Adding an extra highly glazed unheated room – a sunspace or conservatory – to the south side of the house is a further technique for capturing solar energy. The sun warms the air within the space so reducing heat loss from the main building. Such sunspaces can provide a pleasant additional living area available for use for much of the year.

A sunspace or conservatory can be added to an existing building, although the incorporation of such features in the initial design usually achieves a more aesthetically pleasing result.

Estate Layout

To capitalise fully on the potential of PSD it is essential to consider basic estate layout, particularly in the case of domestic buildings. This will require some departure from conventional practice but will not necessarily reduce the dwelling density. For example, the road structure will be influenced by the need to orientate the houses towards the south; the position of the houses within the plots will vary to minimise overshadowing and plot shapes will be governed by orientation requirements. Where mixed dwelling types occur on a site extra care has to be taken to avoid overshadowing. Attention must also be paid to tree planting and other landscaping aspects.

Daylighting

In commercial buildings lighting accounts for a major proportion of the total energy requirement and opportunities exist for achieving substantial savings through the optimum use of daylight. Passive solar building design can produce comfortable levels of even interior lighting without glare and yield significant savings in electricity consumption. Perimeter daylighting uses glass in various locations, usually in the form of windows and clerestories, to allow light to enter the perimeter of the building – baffles and light shelves then distribute the light evenly. Core daylighting systems utilise design features such as roof apertures, light wells, atria and glazed courtyards to bring light to the interior of buildings. Indications are that daylighting features offer dramatic reductions in the energy costs of commercial buildings. Daylit spaces are popular with most building occupants and designs involving atria allow daylighting of deep spaces without the atmospheric exposure of extra windows and walls.

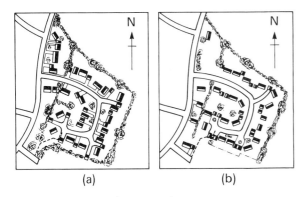

Figure 12.1.2   Housing estate layouts.

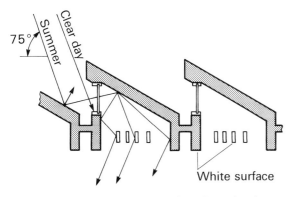

Figure 12.1.3   The pathway of sunlight through a clerestory.

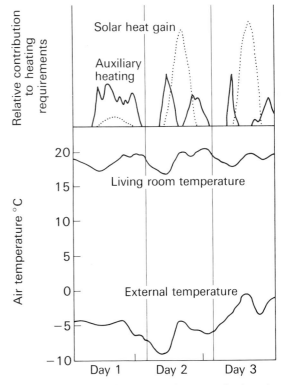

Figure 12.1.4   The relative contributions of solar gain and auxiliary heating in a house, and the external temperature and living room temperature over a three-day period in winter.

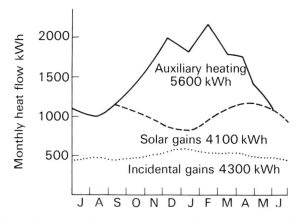

Figure 12.1.5 The relative contributions of auxiliary heating, incidental gain and solar gain in a house over the course of one year.

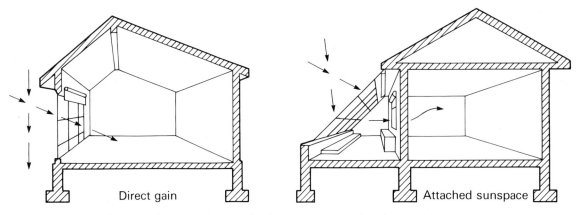

Figure 12.1.6 Schematic diagrams showing (left) direct gain and (right) attached sunspace features.

12 RENEWABLE ENERGY RESOURCES

UNIT 12.2 Biomass energy

Biomass energy is solar energy stored in plants through the process of photosynthesis. It can be used either directly as a fuel (an **energy crop**) or indirectly when waste decomposes.

ENERGY CROPS

Nearly three-quarters of the world's population use wood as their prime fuel, but its use in developed countries is minimal. However, recently in the UK the idea of energy forestry has begun to be developed. The traditional practice of **coppicing** involves cutting back trees and allowing them to resprout. The harvesting of wood from poplar and willow plantations by coppicing every three to five years has already been trialled.

In Brazil sugar cane became an important energy crop in the 1970s. It was used to produce alcohol, which could be used as a petrol substitute (**gasohol**) in specially modified car engines. The massive increase in oil prices in 1974 made gasohol relatively cheap to produce. However in recent years oil prices have stabilised and gasohol has not been such an attractive economic proposition.

ENERGY FROM WASTE

Energy in the form of waste material is thrown away in vast quantities. In the UK it is estimated that the energy equivalent of 25 million tonnes of coal could be retrieved from industrial, domestic and agricultural wastes every year. How can this be achieved? The following possibilities have been researched.

- Burn waste in incineration plants and use the heat produced to generate electricity.
- Compress the waste into brickettes that can be burnt like coal in boilers. This is sometimes known as refuse derived fuel (RDF).
- Hold organic wastes in large tanks in anaerobic conditions (without oxygen). This allows bacteria to decompose the waste, producing methane as a by-product that can be used for electricity generation or as a cooking fuel. The decomposed waste can then be used as fertiliser (see Unit 7.6).
- Collect the gas that is produced by anaerobic bacteria at the centre of rubbish tips. This 'landfill gas' is 60 per cent methane (see box).

ACTIVITY

Using biomass energy

Read the article in the box and answer the following questions.

1 Describe the chain of reactions that converts organic material to landfill gas.

2 Suggest why direct use of landfill gas is the favoured option.

3 Compare the environmental impacts of waste incineration and landfill gas production.

4 What are the advantages of gasohol compared to petrol?

5 Find out which of your local sewage plants use sludge digesters to produce methane. Ask them for details of how they operate and how they make use of the methane produced. What happens to the sewage sludge if it is not digested in this way and how can its disposal harm the environment?

Science, Technology and Society © D. Andrews, 1992

HOW FUEL GAS IS PRODUCED FROM REFUSE TIPS

Modern landfill sites contain large quantities of organic matter; over half the domestic and commercial refuse in the UK is carbohydrate in origin. Consequently, landfill sites represent large "bio-reactors" for the decomposition of this organic matter. Initial microbial activity is high, resulting in rapid depletion of available oxygen and the consequent production of anaerobic conditions such as those which produce "marsh gas". Anaerobic digestion is a multiphase process, mediated by a complex population of micro-organisms. The initial phase involves the degradation of large organic compounds – for instance, those comprising vegetable matter or paper – to simpler substances such as sugars. This is followed by the production of hydrogen, carbon dioxide and fatty acids (eg: acetate) which are the precursors of methane generation in the final stage. The end product of this complicated microbial chain is the biogas commonly known as "landfill gas", which consists of some 40% carbon dioxide (CO_2); 60% methane (CH_4) plus minor constituents.

Landfill gas was first reported in the 1960s in the USA and Germany. Although it was initially treated as a nuisance and abstracted as necessary, the inevitable realisation of its potential energy resource did not take long. The first schemes for exploitation of landfill gas in the UK came in the late 1970s. These involved the firing of brick kilns at London Brick Limited's Stewartby site (work in conjunction with Harwell and funded in part by the Department of Energy) and schemes by Coal Processing Consultants and the GLC at sites such as Aveley, in Essex.

The three main outlets for the landfill gas are:

- direct use
- electricity generation
- gas clean-up and upgrading to higher value fuel.

Direct use is the best option when there is a user in the vicinity of the site and has been successfully demonstrated in kilns and boilers. By the end of 1987 there will be some 14 sites exploiting this use in this country. Electricity generation is the more attractive option for remote sites, and will be in use in five sites in the UK by the end of 1987. At present gas clean-up is hampered by poor economic prospects and technical risks. Current data suggests a national UK potential in the order of 1.3 mtce* per annum from existing sites using present day practices. This is based on a typical yield of 2 therms per tonne of refuse per annum for 12 to 13 years. However, the R&D programme, currently being funded by the Department of Energy and managed by ETSU, will help develop the resource to its full yearly economic potential of 3 mtce. This indicates that landfill gas is one of the most economically attractive sources of renewable energy in this country.

*million tonnes of coal equivalent.

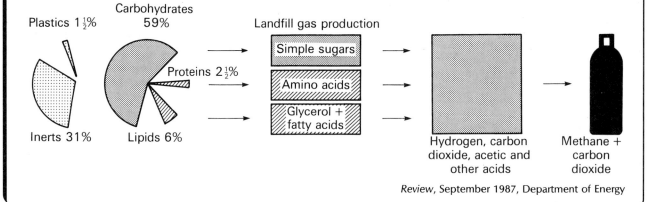

Review, September 1987, Department of Energy

12 RENEWABLE ENERGY RESOURCES

UNIT 12.3 Hot rock power: geothermal energy

The temperature of the rock in the Earth's crust increases steadily with depth. Natural sources of underground water are sometimes heated to boiling point by these hot rocks and forced to the surface as geysers. In parts of the world where geysers are common, such as the USA, Iceland and New Zealand, the steam produced is often used to drive turbines for electricity generation. It is estimated that the total worldwide geothermal generating capacity is over 4700 MW – equivalent to ten coal-fired power stations.

Water below boiling point also comes to the surface at hot springs or aquifers. It can be used directly for heating homes and for industrial purposes. Geothermal hot water is used extensively in Hungary to heat greenhouses.

Hot dry rocks relatively near the surface do not produce geysers but have recently been seen as a potential source of energy. Granite is particularly suitable as it often contains radioactive elements that increase the heat of the rock. To harness this geothermal energy a borehole must first be drilled to a considerable depth through the granite, which is a very hard rock. The rock at the bottom of the boreholes is then fractured using high-pressure water. Water is pumped down the borehole to the hot rocks, where it flows through the rock fissures before returning to the surface under high pressure through a second borehole (Figure 12.3.1).

Since 1973 research has been carried out at the Camborne School of Mines in Cornwall to assess the technical feasibility of **hot dry rock** (HDR) power generation. Government grants have financed three 2 km-deep boreholes. The fractured rock between them has created a heat reservoir that produces up to 30 litres of water per second at 80 °C. It is hoped that commercial investment will be attracted to allow the development of two 6 km-deep boreholes. At this depth the rock temperature reaches 200 °C and the potential for economically viable power generation is much better.

ACTIVITY

The potential of hot rock power

1 Why is granite particularly suitable as a source of geothermal power?

2 Suggest why only certain regions of the UK are suitable for harnessing geothermal power.

3 Why are countries with considerable volcanic activity, such as Iceland, likely to exploit geothermal energy rather than other energy sources?

4 Government financial support for the HDR project in Cornwall was reduced in 1991 because it was not regarded as a viable means of electricity generation. What is your view on this decision?

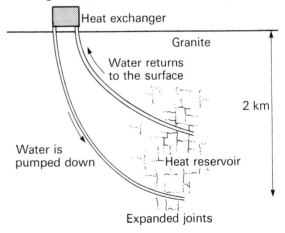

Figure 12.3.1 Schematic diagram of the system of geothermal power production at Camborne in Cornwall.

12 RENEWABLE ENERGY RESOURCES

UNIT 12.4 Tidal power: a feasibility study

The Severn Estuary experiences some of the most extreme tidal conditions in the world. In parts of the estuary the sea level rises and falls by 11 m between high and low tide. This tidal flow could be used to drive turbines inside a barrage built across the estuary. Over several decades there have been many studies of the feasibility of building such a barrage.

The latest study, published in 1989 after two years of careful investigation, firmly supported the idea. However, there were several objections that had to be answered. Would the creation of a barrage seriously damage the environment? Would the electricity it produced be too expensive? Would the local population believe that the tidal barrage would benefit them?

ACTIVITY

The Severn Barrage

Study the excerpts from the Department of Energy information bulletin, which summarises the findings of the feasibility study.

1 Describe and explain the changes that the Severn Barrage is likely to cause in the following:
   a) the greenhouse effect.
   b) the amount of sediment in the basin on the landward side.
   c) the food available to fish and birds.
   d) unemployment in the locality.
   e) the value of property and land.

2 Consider the following local people. Describe what each person's views are likely to be on the building of the Severn Barrage.
   a) The owner of a hotel that is failing to attract tourists.
   b) A person who has recently retired to a quiet spot on the estuary between Barry and Cardiff.
   c) A young person living in Weston-super-Mare who has to drive over the Severn Bridge to visit friends in Barry.
   d) The Secretary of the Yacht Club in Avonmouth.
   e) The owner of a small construction company.

3 In your view, what is:
   a) the strongest argument given in the leaflet in favour of building the Severn Barrage?
   b) the most serious detrimental effect that the Barrage might have?

4 Find out more about the latest progress of the Severn Barrage scheme by writing to:

   Renewable Energy Enquiries Bureau
   ETSU
   Building 156
   Harwell Laboratory
   Oxfordshire
   OX11 0RA.

Department of Energy Information Bulletin

ISSUE 7 NOVEMBER 1989

SEVERN BARRAGE DEVELOPMENT PROJECT

SEVERN BARRAGE DEVELOPMENT PROJECT

The most detailed study to date of the possibility of harnessing the extreme tidal conditions in the Severn Estuary to generate electricity is now complete.

The Severn Tidal Power Group's findings of the study are summarised in this Bulletin.

Summary of Findings

Implementation Programme

The present studies have not included any work on organisation and financing but it has been estimated from previous studies that the pre-construction stage of the project would take five years and cost £190 million. The construction programme envisages a production rate of 44 turbine generators per year, seven years to the closure of the barrage and the first commercial power generation when 144 turbines would be in operation, and a further two years to completion and full output.

Electricity Production and Costs

The average annual output of electrical energy would be about 17 terawatt hours, equivalent to 7% of the present electricity consumption in England and Wales and result in the saving of the equivalent of 8 million tonnes of coal annually. Thus electricity generated by this means would reduce carbon dioxide emissions by up to 17.6 million tonnes per year, contributing to the long-term strategy of minimising the greenhouse effect.

The cost of electricity at the point of transmission from the barrage would range from 1.7 pence per kWh at 2% discount rate to 8.2 pence per kWh at 10% discount rate.

A design life of 120 years has been allowed for, but the life is likely to be much longer with normal maintenance and equipment replacement.

Environmental Effects

The effects of a barrage scheme on the environment would be a major consideration in determining its acceptability. The environmental assessment work undertaken for this study, though not exhaustive, is as wide-ranging as any carried out to date on a British estuary.

The key to any change in the environment is the predicted change in the water levels and currents in the estuary. Mathematical modelling has shown that landward of the barrage mean spring tide levels would be reduced by half a metre or less, and low tide levels would generally only fall to about mean sea level. Outside the barrage, tide levels would be much less affected. Landward of the barrage currents would be, at most, half those occurring at present and seawards they would also be reduced; this would cause a major change in the movement of sediment in the estuary and a reduction in the amount of sediment suspended in the water.

The changes in the deposition of sediment landward of the barrage are not expected to have a significant effect on barrage operation but there may be some accretion in the rivers discharging into the estuary and this would need further quantification. The reduction in the amount of sediment suspended in the water is expected to lead to an increase in primary biological production in the estuary and hence to an increase in the food available to fish and birds. Water pollution is not expected to increase generally over present levels.

➡

Further studies of the interactions between the flora and fauna of the estuary will be necessary, but those carried out to date indicate that overall the number of wading birds and wildfowl is as likely to increase as to decrease, while the diversity of species would almost certainly increase following barrage construction. Fish and shrimp might also increase in both population and size and range of species, but increased understanding of damage done to them by turbines, together with further studies of possible designs for fish passes and deterrents, is still needed.

The study has also shown that the works needed to deal with changes in land drainage by increased water levels landward and seaward of the barrage are likely to be localised and of minor cost. A particular benefit which might accrue with the barrage in operation is the protection of a long length of coastline landward of the barrage from high tidal surges and rising sea levels which could pose a significant threat in the 21st century. On the negative side, a possible lessening of bank stability might occur due to higher mean water levels, but the net costs seem unlikely to be significant.

Regional and National Effects

Economic studies have indicated that the building of the barrage would result in a boost to the economy of the region stemming from the construction and operation of the barrage, tourism and recreation, industrial and commercial property development, road transport, and housing and infrastructure. In addition other parts of the country, particularly the manufacturing regions, would also benefit from the construction work.

Total employment on the barrage construction would amount to some 200,000 man-years with a peak of about 35,000 jobs in the third year of construction. About half of these would relate to the local region. Following construction of the barrage it is estimated that there would be a build-up to an additional 10,000–40,000 permanent jobs in the region – 1,000 resulting from barrage-operation, 2,000 from increased tourism and the remainder from industrial, commercial and other activities. Substantial additional income would be generated and retained within the region and land and property values would be expected to increase.

On the basis of present studies, the effect of the barrage on local ports may prove to appear to be broadly neutral, but more detailed modelling of shipping movements will be necessary to establish the true position.

The economy of the South West, and more particularly of South Wales, would be materially assisted by the construction of a dual carriageway across the estuary some 45 km seaward of the M4 crossing. The transportation opportunities thus generated would represent a national economic gain if the road across the barrage were to be connected to the motorway network with improved links to the A303 route.

In Conclusion

The study has consolidated previous work and confirmed that the Severn Barrage is a feasible project, well within the scope of existing technology. Moreover, it is likely to have a very long life. If renewable energy sources are to be utilised to increase diversification of electricity generation and to reduce pollution, it remains the largest single project which could make a significant contribution on a reasonable time-scale. The study suggests that the construction and operation of the barrage would result in a major boost to the economy of the region.

Work on environmental issues has lessened uncertainties and identified a number of possible benefits that would accrue with the barrage in operation, but much additional work still needs to be done before a full environmental assessment is completed.

The Barrage Project

The 16 km (10 miles) long barrage would be constructed mainly of reinforced concrete caissons made at shore facilities and floated to the site. There would be embankments adjacent to the Welsh and English coasts and near Steep Holm island. A dual carriageway public road across the barrage would link to the main highway system on both shores.

Power would be transmitted by 400 kV cables from the barrage to sub-stations at both sides of the estuary and delivered into the national grid.

In the reference scheme which forms the basis for the cost estimate, large locks are included which would be capable of passing, at most states of the tide, the largest ships (about 70,000 dwt) currently trading to the basin. Small craft locks would also be provided near each shore.

Basic data for the reference scheme is as follows:

Number of turbine generators	216
Diameter of turbines	9.0 m
Operating speed of turbines	50 rpm
Turbine generator rating	40 MW
Installed capacity	8640 MW
Number of sluices, various sizes	166
Total clear area of sluice passages	35000 m^2
Average annual energy output (10^9 kWh = 1 TWh)	17 TWh
Operational Mode: Ebb generation with flood pumping	
Length of barrage, total	15.9 km
including: powerhouse caissons	4.3 km
sluice caissons	4.1 km
other caissons	3.9 km
embankments	3.6 km
Area of enclosed basin at mean sea level	480 km^2

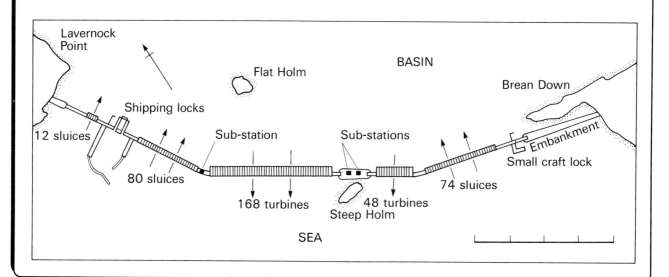

12 RENEWABLE ENERGY RESOURCES

UNIT 12.5 # Hydropower: large-scale and small-scale

Power production on a large scale will inevitably have a more noticeable environmental impact than small-scale systems. This applies to renewable energy technologies as much as to power generation from non-renewable resources.

Building a dam across a river valley holds back a large expanse of water in a reservoir and provides the potential for hydroelectric power. The dam may have other benefits, such as the ability to control flooding below it. The reservoir it creates can open up opportunities for many leisure activities such as sailing and fishing. The stored water can be used in homes and in industry as well as for irrigating crops threatened by drought. However the environmental impact of huge dams can be catastrophic. The Itaipu dam on the Brazil–Paraguay border formed a lake 184 km long and displaced 42 000 people, and the Aswan Dam on the River Nile has created a number of unforeseen problems (see box).

Hydropower that does not involve massive dams has been commonplace for centuries. The waterwheel was the main power source for industry before the era of steam power. Waterwheels rely on a natural steep drop as water flows downhill. Modern, small-scale hydropower schemes enclose the water flow in a pipeline with a turbine at its lower end. They are particularly useful in remote, hilly regions that do not have access to a national grid. China has installed 87 000 of these schemes since 1968. Between them they produce 5000 MW of electricity, the equivalent production of ten coal-fired power stations. The introduction of electrical power to isolated villages can help to boost local industries, as well as fulfil domestic cooking and lighting needs.

Aswan: The Unexpected Problems

An important conference in Cairo ended today on a gloomy note. The conference, convened for scientists to exchange information about the problems caused by the Aswan Dam, revealed a worrying future scenario.

The scientists acknowledged that the dam has given Egypt many benefits. The Nile Delta no longer floods every year and hydroelectric power from the dam provides half the country's electricity needs. But these advantages are set against a growing number of environmental and economic problems.

The large quantities of water flowing on to the land from the new irrigation can cause serious waterlogging, leading to soil salinisation. This occurs when water evaporates rapidly, leaving behind a white salty crust on the surface, which makes it difficult to grow crops.

Over a million tonnes of silt, clay and sand, which used to fertilise the downstream fields during periods of flooding, are now silting up Lake Nasser. Farmers have therefore had to increase their spending on expensive imports of chemical fertilisers. The Cairo brick-making industry has also been devastated because it no longer has a ready supply of clay. The most disconcerting news at the conference came from a scientist who claimed that at the current rate of accumulation Lake Nasser would be entirely filled with sediment within the next 30 years.

ACTIVITY

Hydropower problems

1 Give three major advantages of hydropower compared to the generation of electricity in a coal-fired power station.

2 The expected life of a dam is over a hundred years, but its useful life for electricity generation will be less than this. Explain why.

3 How do you think the scientist mentioned in the article in the box made the calculation about the time taken for the lake behind the Aswan Dam to completely fill up with silt?

4 Debate this motion in groups or as a class.
 'On balance the environmental and economic costs of large-scale hydropower schemes outweigh the environmental and economic benefits.'

5 Discuss the advantages of small-scale as opposed to large-scale hydropower.

12 RENEWABLE ENERGY RESOURCES

UNIT 12.6 The potential of wind power

Some of the solar radiation heating the Earth's atmosphere is converted into wind energy. Air masses in different parts of the atmosphere are heated unevenly by the differing intensities of the Sun's rays. Warmer air rises, allowing cooler air to rush in. This rapid movement of air represents a massive untapped source of energy. Windmills, designed to convert this energy source to motive power, first appeared in China around 2000 BC. The heyday of wind power in the UK was in the late eighteenth century, when there were 10 000 windmills in operation.

Wind has an important advantage as a source of energy: the power produced by wind is proportional to the cube of the wind speed. Thus if wind speed doubles, the power available increases eight-fold. However against this, wind has several disadvantages. It can be erratic and unpredictable, as sailors becalmed for days in mid-ocean can testify. Its speed will only reach exploitable levels in certain geographical spots.

A wind turbine or aerogenerator converts wind energy into electrical energy. As wind flows over and turns the rotor blades, the rotational movement drives an electrical generator. The largest wind turbine in the world, with a 3 MW capability, is located on Burgar Hill on the island of Orkney. It has a two-bladed horizontal axis design measuring 60 m in diameter and can generate 9 million kWh annually, enough for 2000 homes.

- Total UK annual electricity consumption = 250 000 GWh
- 1 GWh = 1×10^6 kWh
- It is estimated that five 3 MW aerogenerators would require a land site of one square kilometre (1 km^2).

ACTIVITY

The impact of wind power

Use the information given and the facts in the box to answer the following questions.

1. What is the load factor (see Unit 11.3) of the Burgar Hill aerogenerator?

2. Some scientists believe that 10 per cent of our current electricity demand could be met by wind power. How many 3 MW wind turbines would be required to meet this demand?

3. Calculate what land area would be taken up by this number of generators.

4. What do you feel to be
   a) the strongest argument in favour
   b) the strongest argument against
   the introduction of wind power on this scale?

5. Remote, exposed hilltops would be an ideal location for wind forms. But this is precisely where environmental objections would be strongest. Why?

6. What technical and economic problems might be encountered in siting wind turbines on offshore platforms?

7. Study the two designs for wind turbines currently under development, shown in Figures 12.6.1 and 12.6.2. Which do you consider to have the greater potential?

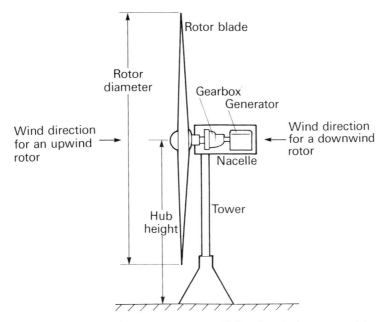

Figure 12.6.1 Horizontal axis wind turbine. The rotor's axis is parallel to the wind stream and the ground. The generator is at the top of the tower.

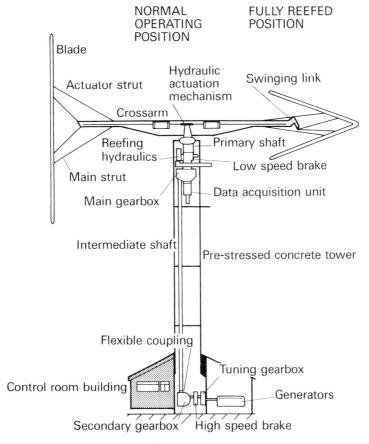

Figure 12.6.2 Straight blade vertical axis wind turbine. The rotor's axis is at right angles to the wind stream and the ground. At high wind speeds the blades are 'reefed' into an arrowhead shape to regulate the power output.

SCIENCE TECHNOLOGY AND SOCIETY

12 RENEWABLE ENERGY RESOURCES

UNIT 12.7 Wind power in the developing world

Large wind turbines of up to 3 MW are designed to feed electricity into a national grid system. This allows fluctuations in supply to be evened out by co-ordinating electricity generation over a large number of power stations. Many developing countries do not have a grid system so an alternative approach to wind power is required. 'Stand-alone' wind power systems consist of a single wind turbine connected to batteries that store electric charge for times when the wind speed drops. When the wind speed is low the batteries can continue to provide electricity. In this unit you will explore the technical problem of setting up a wind power system to match the electricity demand.

An engineer helping a village to set up a 'stand-alone' wind generator uses the equation below.

Table 12.7.1 Energy requirements of electrical equipment.

Application	Use	Mean daily energy demand (kWh)
Water pumping	Irrigation Drinking	3.0
Refrigeration	Food storage Vaccine preservation	1.0
Lighting	Homes Work areas	0.25 per 60 W bulb
Communication	Televisions Telephones Radio transmitters	0.5

Mean daily energy output from the wind turbine (kWh) = $0.0048 \times a \times v^3$

where

a = area of sweep of rotor blades (m²)
v = mean wind speed (ms⁻¹)

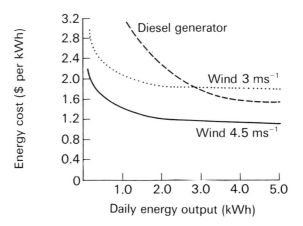

Source: Intermediate Technology Group, Rugby

Figure 12.7.1 A comparison of the costs of different methods of electricity generation for different output levels.

ACTIVITY

Wind power feasibility study

1. Calculate the daily energy output from wind generators with the following characteristics.
   a) 1 m radius rotor blade at a mean wind speed of 2 ms^{-1}.
   b) 1 m radius rotor blade at a mean wind speed of 3 ms^{-1}.
   c) 1 m radius rotor blade at a mean wind speed of 4 ms^{-1}.

2. Using the results from 1, find out what happens to the energy output when the wind speed doubles.

3. The village has a small medical centre. It needs electricity to power six 60 W light bulbs, a radio transmitter and a vaccine refrigerator. Estimate the total daily energy demand of the medical centre using the data in Table 12.7.1.

4. A wind generator with rotor blades 2 m in diameter is available. If the mean wind speed is 3 ms^{-1}, how many of these generators would be required to meet the medical centre's energy demand?

5. a) If a single generator was preferred, what would be the required rotor sweep area?
   b) What would be the diameter of these rotor blades?

6. Study the graph in Figure 12.7.1 and use it to answer these questions.
   a) In a village where the mean wind speed is 3 ms^{-1}, a home's daily energy demand is 2.5 kWh. What would be more economical, a wind generator or a diesel generator?
   b) In the same village, a larger home has an energy demand of 4.0 kWh. Which method of power generation would be more economical in this case?
   c) In general terms, what does this graph tell you about the economics of wind power as an energy source in developing countries?

12 RENEWABLE ENERGY RESOURCES

UNIT 12.8 Wind and wave power under attack

Using the power of wind and waves to generate electricity is a controversial issue. The technology to harness the vast amount of energy in the oceans' waves is in the early stages of development (see Box 1). Wind power is well-established in California, but its potential is uncertain in the rest of the world. Scientists have widely differing views concerning the viability of utilising both these energy sources.

A dramatic storm in January 1989 raised new doubts. A wave power station built on to a cliff face of Norway's North Sea coastline was smashed into the sea by 20-metre waves. The storm also wrecked a wind turbine on the Shetland Islands, tearing off an 8-metre blade and throwing it 80 metres on to a power line.

ACTIVITY

Views on wind and wave power

The effects of the storm in January 1989 provoked differing responses. An editorial from the *New Scientist* magazine of 14 January 1989 is reprinted in Box 2. Two letters from the following week's issue responded to this editorial. Excerpts from these letters are given in Box 3. Study this material and use your knowledge from previous units to discuss the following questions.

1. a) Where is Tonga?
   b) Why is it particularly suitable for a wave power plant?

2. a) Why does the writer of the editorial argue that the Tonga people have 'landed themselves in 5 million pounds-worth of trouble'?
   b) Do you agree?

3. Explain why renewable energy and nuclear power might be considered 'equally beneficial' in solving the problem of the greenhouse effect.

4. a) The editorial argues that the advocates of alternative energy are 'high on emotion and low on reality'. What does this mean?
   b) What evidence does the writer put forward to support his view?
   c) Find evidence from Douglas Dale's letter that contradicts this view.

5. a) How does the potential impact of a storm-damaged wind turbine or wave power plant compare with the consequences of an accident at a nuclear reactor or oil production platform?
   b) Does this comparison provide support for renewable energy?

6. a) What is meant by a power plant's 'load factor'? (See Unit 11.3.)
   b) How does the Californian wind turbine's load factor compare with other means of electricity generation?
   c) Suggest reasons for this difference.

7. In what way might the accident at the Norwegian power plant be useful to researchers developing wave power?

8. Why does the *New Scientist* editorial maintain that low oil prices have a detrimental effect on energy conservation?

9. Suggest reasons why inshore wave power devices are regarded as more promising than devices designed to operate in the open sea.

Wave power devices

The awesome power of storms at sea are well-known to any sailor or offshore oil rig worker. It is not surprising that designing a device that can survive these conditions while generating a useful amount of electricity has proved to be a daunting task. Offshore devices such as the massive nodding duck named after its inventor Stephen Salter are still at the early stages of development (Figure 12.8.1). Inshore devices appear more promising. These devices are built in natural rock gullies on exposed shorelines and operate on the principle of the oscillating water column (OWC). A specially constructed concrete chamber captures water as it flows in and out of the gully. The air enclosed above the water column is forced through an air turbine which drives an electricity generator (see Figure 12.8.2).

Since the 1970s the Japanese have used devices like these to power lighthouses and in 1985 a 500 kW OWC wave power station was installed on the Norwegian coast north of Bergen. In 1990 a prototype OWC device incorporating the newly developed Wells turbine was successfully trialled on the Island of Islay off the coast of Scotland. This turbine, which was developed by the scientist Alan Wells, always revolves in the same direction regardless of the direction of the airflow. As a result electricity is generated continuously as the water column oscillates.

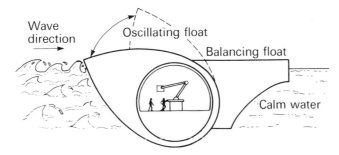

Figure 12.8.1   The design of Salter's duck for harnessing wave power. A full-size prototype is yet to be built.

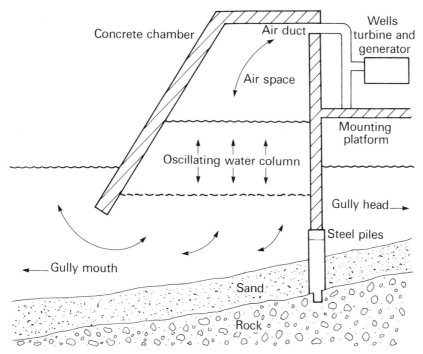

Figure 12.8.2   Schematic diagram of the operation of the OWC system for harnessing wave power.

Science, Technology and Society © D. Andrews, 1992

② Wave in and drowning

If the people of Tonga suffer from the same severe storms that afflict countries around the North Sea, they may have landed themselves with 5 million pounds-worth of trouble. This trouble will come in the shape of a new wave-power plant. Their plant, said to be the world's first commercial wave-power system, will be built by the same company that put up the Norwegian wave-power station that came to grief during a severe storm over the New Year weekend.

It would, of course, be daft to reject renewable energy on the strength of one accident – or two to be more precise, because a wind turbine fell over a few days earlier – but it would be equally silly to suggest that waves, wind, tides and other sources of renewable energy are without their drawbacks.

Advocates of these alternatives to fossil fuel and nuclear reactors are often high on emotion and low on reality. So when someone has the temerity to challenge their technology on the grounds of poor economics they run around crying 'fix'. The environmentalists further dilute their credibility when they hold up renewable energy sources as 'The Answer' to the greenhouse effect while deriding anyone who dares to suggest that nuclear power stations might be equally beneficial on the same front. Storms and carbon dioxide prove that little beyond the fact that life isn't black and white and that nothing is perfect on the energy front. Even the CEGB now concedes that nuclear power is not the economic walkover that it has been presenting it as in recent years. Just about the only thing that you can say about energy is that it makes sense to use it as wisely as possible. Unfortunately, oil prices are on the floor and Britain at least seems to have given up on pushing energy conservation.

New Scientist, 14 January 1989

③

Making waves

It was a little irritating to see your Editorial 'Wave in and drowning' which gives the impression that those who support further work on wave energy are a little soft in the head. We are not. We only claim that:

1. There is a vast amount of energy in waves on some coastlines.
2. Britain has some potentially excellent sites.
3. The OWC is likely to produce electricity at or near the costs of production by existing methods.
4. It will certainly be economic for some remote locations and quite possibly for general supply.
5. The public has been misled about the cost of nuclear power, as has become evident during the past 18 months.
6. Wave energy would cause no pollution.
7. A prototype three-column station could have been in production three or four years ago at a cost estimated then of 15 million pounds – crumbs from the nuclear table.

It is as sensible to rubbish wave energy because of the Norwegian accident as it would have been to stop railway development on account of the Tay Bridge disaster.

Douglas Dale
Stoke-on-Trent

Problems with prototypes

Your perceptive editorial on the exaggerated hopes for new energy technologies is welcome and it is to be hoped that it has been noted by Britain's Department of Energy which has made absurd claims on behalf of wind turbine generators in California, ignoring their load factor of only 9.1 per cent.

That said, it is, I hope, fair to add that established energy sources also felt the strain in the exceptionally severe weather: a 200 000 tonne supertanker used as a floating oil collection platform was torn loose from its moorings ... The oil industry has had time (and money) to learn about the North Sea and if it finds the weather overwhelming newcomers to the scene may surely be forgiven.

Without complacency, it may be remarked that the Norwegian wave energy unit was a prototype and one could almost say that this is what prototypes are for. Sir Hermann Bondi, when he was Chief Scientist, was always encouraging the wave energy researchers to hurry on down into the water on the grounds that the sea always had a trick up its sleeve and this could be discovered only on what he called 'a green sea site'. The Norwegians did not allow themselves to be confined to the laboratory and they have proved him right. And they have, as they themselves say, now had the experience and have learnt from it.

David Rose
London SE5

New Scientist, 28 January 1989

12 RENEWABLE ENERGY RESOURCES

UNIT 12.9 The future of renewable energy technologies

In the late 1990s it is likely that renewable energy sources will begin to make an important contribution to the UK's energy needs. In 1989 the UK government published a proposal that a minimum of 600 MW of electricity generating capacity should come from renewable energy technologies by the year 2000. The figures shown in Tables 12.9.1 and 12.9.2 were produced by the Department of Energy in 1989 to indicate the technical potential and the estimated contribution of various renewable energy technologies to overall energy requirements in 2025.

Table 12.9.1 Technical potential and estimated contribution of renewable electricity producers.

Technology	Technical potential TW h/year*	Estimated contribution TW h/year
Onshore wind power	45	0–30
Offshore wind power	140	uncertain
Tidal	54	0–28
Geothermal	210	0–10
Wave	50	0–0.2
Small-scale hydroelectric	2	0.3–0.7

*1 TW = 10^{12} W

Table 12.9.2 Technical potential and estimated contribution of renewable heat producers.

Technology	Technical potential Mtce/year*	Estimated contribution Mtce/year
Passive solar	8–14	1–2
Waste as fuel	22	3–10
Energy forestry	20	1–5

*Megatonnes coal equivalent per year.

In 1989 the Department of Energy produced an assessment of the likely future contribution of renewable energy to the UK's overall energy requirements (Table 12.9.3). This was to help determine the amount of financial support the government would give to the research and development of the various renewable energy technologies. To assist this process each one was assigned to one of three broad categories:

- **Economically attractive** – technologies that are cost-effective in some markets at present and seem likely to make a significant contribution to UK energy supplies. They should be encouraged to a commercial scale by the 1990s.
- **Promising but uncertain** – technologies likely to become competitive if predicted costs can be achieved or if fuel prices rise in real terms. They should be given further support to bring about improved performance and lower costs.
- **Long shot** – technologies that would be cost-effective only in the unlikely event of a dramatic improvement in costs or a sharp increase in fuel prices. They should be given only minimal support.

ACTIVITY

Renewable energy policy

1 Explain why the three renewable energy technologies in Table 12.9.2 have been classified as heat producers rather than electricity producers.

2 What are the environmental factors that explain why the technical potential of renewable energy technology is more than its estimated contribution to UK energy supplies?

3 How many large fossil-fuel power stations does it currently take to produce the 600 MW of electricity generating capacity that is to come from renewable energy technologies by the year 2000?

4 The present UK total electricity consumption is 290 TW h/year.
   a) If the maximum estimated contribution is fulfilled, what percentage of this will be produced by renewable energy technologies?
   b) Discuss this estimated percentage. Should it be higher? If so, how would it be achieved?

Science, Technology and Society © D. Andrews, 1992

Table 12.9.3 The likely future contribution of renewable energy to the UK's energy requirements.

Energy	Category
Passive solar design	Economically attractive
Active solar heating	Long shot
Photovoltaics	Long shot
Combustion of dry wastes	Economically attractive
Anaerobic digestion, e.g. landfill gas	Economically attractive
Thermal processing of dry wastes	Promising but uncertain
Energy forestry	Promising but uncertain
Wind energy – *on land*	Promising but uncertain
Wind energy – *offshore*	Long shot
Tidal energy	Promising but uncertain
Inshore wave energy	Promising but uncertain
Small-scale low head hydro power	Promising but uncertain
Large-scale offshore wave energy	Long shot
Geothermal hot dry rocks	Promising but uncertain
Geothermal aquifers	Long shot

ACTIVITY

Economics of renewable energy

1 Why is the cost of fossil fuel important to an assessment of the prospects for renewable energy?

2 Some governments provide subsidies to ensure that renewable energy technology thrives even if it is not commercially viable. Californian wind farms are one example of this. Why is this done? What are your views on this policy?

3 Study the assessments shown in Table 12.9.3. How far do you agree with them?

13 NUCLEAR POWER

UNIT 13.1 What is radioactivity?

Radioactivity conjures up many fears. It is something that we cannot see, smell, touch or hear, yet in high doses it can have harmful effects on living tissue. This unit gives some of the facts about radioactivity and shows how some radioactive substances can be useful to us.

WHAT ARE RADIOISOTOPES?

A **radioisotope** is the form of an element that releases radioactivity. To understand how this happens you need to appreciate how atoms are constructed.

A single atom is not just a tiny solid ball of matter. It can be broken down further to simpler particles. Clustering together in the atomic nucleus are **protons**, with a positive charge, and **neutrons**, with no charge. Much lighter, negatively charged **electrons** orbit the nucleus at high speed. The number of protons gives the atom its atomic number. This is equal to the number of negatively charged electrons that orbit the nucleus. The **atomic number** of a given element is constant: for example, carbon has an atomic number of 6. An atom's **atomic mass** is the number of neutrons plus the number of protons. However, it is possible for an atom of the same element to have different atomic masses. These are the element's **isotopes**. Carbon exists as two isotopes: carbon-12 and carbon-14.

Certain isotopes are unstable and are known as radioisotopes. The nuclei of these radioisotopes rearrange themselves in an attempt to become stable. In doing so they give off particles or electromagnetic rays. In other words they emit radiation. There are three common types of radiation: alpha, beta and gamma. All of these types of radiation can damage human cells by altering the structure of the DNA molecules (see Unit 4.4).

THE THREE TYPES OF RADIATION

- **Alpha particles** consist of two neutrons and two protons. They are relatively heavy and slow moving and will pass through a thin sheet of paper. If an isotope that emits alpha particles enters the body through breathing or swallowing in the form of dust, it will be 20 times more damaging than an isotope emitting beta particles. However, outside the body alpha particles are relatively harmless since they rarely penetrate the dead outer layer of the skin (epidermis).

- **Beta particles** are high energy, fast-moving electrons that have a greater penetrating power than alpha particles. They can penetrate several millimetres of aluminium and will pass through the surface skin layers. Someone exposed to high doses of beta radiation will suffer severe burns. Such 'beta burns' were common among the firefighters who tackled the blazing Chernobyl reactor (see Unit 13.5).

- **Gamma rays and X-rays** are electromagnetic radiations similar to light and radiowaves. They have a very high energy value and can penetrate deep into human tissue, damaging the DNA molecules, causing genetic mutations and increasing the risk of cancer. The penetrating power of this type of radiation is so high that it can pass through a 2.5 cm thick aluminium sheet. A barrier of 2 m of concrete or 2 cm of lead is required to stop it.

RADIOACTIVE DECAY

A radioactive material will continue to emit radiation until all its atoms reach a stable state. As time passes, the number of unstable atoms gradually decreases and the amount of radiation released will fall. The process is called **radioactive decay**. The rate of decay is expressed as the **half-life** of a particular isotope. This is the time taken for the level of radioactivity to decrease to half its original value. The half-life value for some important isotopes is shown in Table 13.1.1.

Table 13.1.1　The half-lives and type of radiation emitted by some radioisotopes.

Isotope	Half-life	Radiation	Notes
Carbon-14	5730 years	Beta	Used in fossil-dating
Strontium-90	28 years	Beta	Absorbed into bone structure
Iodine-131	8 days	Beta, gamma	Major isotopes in Chernobyl dust cloud
Caesium-134	2 years	Beta, gamma	
Caesium-137	30 years	Beta, gamma	
Radon-220	55 seconds	Alpha	Radioactive gas
Uranium-238	4500 million years	Alpha	Most common isotope in mined uranium
Uranium-235	710 million years	Alpha	Thermal nuclear reactor fuel
Plutonium-239	24 400 years	Alpha	Breeder nuclear reactor fuel
Neptunium-237	2 200 000 years	Alpha	Decay product of fission of Uranium-235

ACTIVITY

1. How long will it take for the radioactivity in the following radioisotopes to drop to a quarter of its original value:
   a) Uranium-238
   b) Caesium-134
   c) Radon-220?

2. Use the information given about alpha radiation to explain the following:
   a) It is safe for miners to handle natural uranium ore without gloves.
   b) Uranium miners have a higher than average risk of lung cancer.

3. The amount of a particular isotope in a fossil can be used to date it. Which isotope is this, and why could it not be used to date a fossil that is over five million years old?

4. After a nuclear accident iodine tablets can be taken as a precautionary measure to block the absorption of iodine-131. Suggest why people could stop taking them after four weeks.

5. Explain the difference between:
   a) an isotope and a radioisotope
   b) a proton and a neutron
   c) alpha radiation and beta radiation.

13 NUCLEAR POWER

UNIT 13.2 How do nuclear power stations work?

Nuclear power has been the subject of considerable controversy. When it was first introduced in the 1950s some people envisaged that nuclear power stations would produce electricity that would be 'too cheap to meter'. This was a wildly over-optimistic claim and it now appears that the cost of electricity from UK nuclear power stations is slightly greater than that from coal-fired power stations.

The environmental impact of nuclear power is also fiercely debated. While it produces no carbon dioxide and therefore does not contribute to the greenhouse effect (see Unit 11.9) it does produce radioactive waste that must either be reprocessed or disposed of. The risks of an explosion at a nuclear power plant releasing huge quantities of radioactive fallout must also be considered. This unit investigates the operation of nuclear power stations so that your view of the safety question is based on firm information.

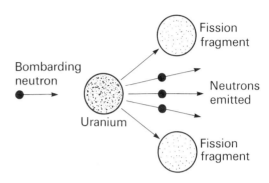

Figure 13.2.1 Nuclear fission of a uranium atom to produce two fission fragments, three neutrons and energy.

CHAIN REACTIONS AND NUCLEAR POWER

Uranium is a naturally occurring mineral that is mined in many parts of the world. After the uranium ore is purified and formed into fuel rods for the nuclear reactor it is almost 99.3 per cent made up of the isotope uranium-238 and 0.72 per cent of uranium-235. When bombarded with a neutron, the relatively large and unstable uranium-235 atom splits into two approximately equal portions (see Figure 13.2.1). At the same time, two or three fast-moving neutrons are released. This is nuclear **fission**, or splitting the atom, and was first discovered in 1938 (see Unit 4.2). When the neutrons from the fission reaction collide with other uranium-235 atoms, these also split and release more neutrons. These reactions multiply rapidly and form a chain reaction releasing vast amounts of energy (see Figure 13.2.2). Some neutrons are absorbed by the uranium-238, which decays through several steps to form plutonium-239. After several years in the reactor the uranium in the fuel rod is depleted, and the rod is then replaced. Spent fuel rods are rich in plutonium-239 and still highly reactive. They must be disposed of or reprocessed (see Unit 13.3).

An uncontrolled chain reaction occurs when a sufficient mass of pure uranium-235, the **critical mass**, is

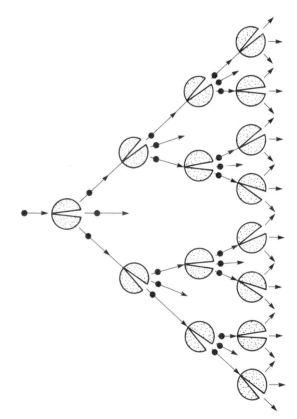

Figure 13.2.2 Chain reaction of uranium fission. Each fission releases neutrons, which cause further fission reactions.

formed. This is the principle on which the first atomic bomb was built. Although a nuclear reactor contains more uranium-235 than an atomic bomb, it cannot explode like one. This is because of the relatively low proportion of uranium-235 in the fuel rods (up to 2.3 per cent) and their distribution over the large area of the reactor. Nevertheless a conventional explosion and

Science, Technology and Society © D. Andrews, 1992

fire is possible within a reactor and this can release radioactive fallout over a wide geographical area, as occurred after the disaster at the Chernobyl nuclear power station in 1986.

Nuclear power became possible when we discovered how to control the chain reaction so that a steady release of heat is achieved. Within the core of the nuclear reactor the uranium fuel rods must be separated by another material (the moderator) that slows down the neutrons and prevents an uncontrolled chain reaction. Above the core are adjustable boron rods that can be raised and lowered to control its temperature. If core temperature is too hot the boron rods are lowered which absorb extra neutrons, the chain reaction slows down and the temperature drops. If severe overheating occurs all the boron rods are lowered fully into the core and the chain reaction is brought to a complete halt.

A coolant circulates through the core, transferring the heat to an exchanger, which raises steam to drive a turbine. A loss of coolant can lead to disastrous overheating and a resultant melt-down of the reactor core, as happened in the Chernobyl nuclear reactor accident in 1986 (see Unit 13.5). In the UK, most nuclear reactors use carbon dioxide gas as the coolant. The first Magnox reactors were built in the late 1950s and used un-enriched uranium. The more efficient advanced gas-cooled reactors (AGRs) were introduced in the 1970s. They use fuel rods enriched to 2.3 per cent uranium-235 and operate at higher temperatures and pressures. Worldwide the most popular reactors use pressurised water as a coolant (pressurised water reactors, PWRs). Figures 13.2.3 and 13.2.4 show the basic design of these two types of reactors.

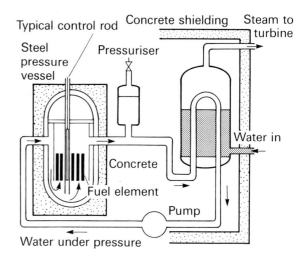

Figure 13.2.4 Schematic diagram of a pressurised water reactor.

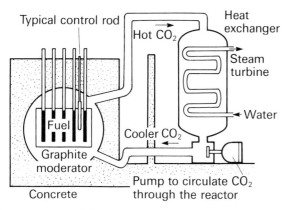

Figure 13.2.3 Schematic diagram of a Magnox nuclear reactor.

ACTIVITY

1 Explain the main differences between the Magnox reactor and the PWR.

2 Find out the current position on the plans for PWRs in the UK by writing to Nuclear Electric.

3 Discuss the safety factors that are built into the design and operation of nuclear power stations. How far are you reassured by them?

4 Anti nuclear groups, such as Greenpeace and Friends of the Earth, argue that nuclear power stations pose too great a risk and they should all be shut down. Find out about their concerns and discuss this proposal.

13 NUCLEAR POWER

UNIT 13.3 Reprocessing and breeder reactors

When uranium fuel rods have been removed from nuclear reactors they can either be disposed of or reprocessed. Disposal is surrounded by a number of technical and political issues, which are explored in Unit 13.4. Reprocessing involves the separation of the 40 or more radioisotopes produced by uranium fission within the fuel rods. The most important of these isotopes is plutonium-239, which is an essential component in nuclear warheads. Plutonium can also be used as fuel in another type of nuclear power station currently at the prototype stage of development in the UK – the breeder reactor. The link between reprocessing and breeder reactors is illustrated in Figure 13.3.1.

Uranium is a non-renewable fuel. At the present rate of consumption the world's nuclear fission reactors will use up the earth's uranium resources in 125 years. However a breeder reactor can produce as much fuel as it consumes. The reactor core of plutonium is surrounded by a blanket of uranium-238 10 cm thick. Fast neutrons from the fission of plutonium hit the uranium-238 and convert it into more plutonium.

The great advantage of the breeder reactor is that its fuel comes from the spent fuel rods that are currently stockpiled as high level nuclear waste. If the 20 000 tonnes of this waste now stored in the UK were used in this way it would produce more energy than the entire 300 years' supply left in the UK coal reserves.

However there are many technical and safety problems that must be overcome before breeder reactors can progress beyond the prototype stage. The intensity of the fission reaction is a major problem in the design of the reactor's cooling system. Instead of gas or water as a coolant, liquid metals such as pure sodium, with a higher heat capacity, must be used. If the molten sodium comes into contact with water or steam then it will explode. This makes breeder reactors considerably more hazardous and expensive than ordinary nuclear reactors.

Britain's prototype breeder reactor at Dounreay in the far north of Scotland has been operating since 1959. However no full-scale commercial reactor has yet been built in the UK and there are only five operating in other countries.

Another argument against breeder reactors is that the uranium fuel rods need to be reprocessed to separate out the plutonium. In this form the plutonium could be stolen by terrorist groups to manufacture their own bombs. If the spent fuel rods are stored rather than reprocessed the plutonium is dispersed amongst the uranium and is inaccessible for this purpose.

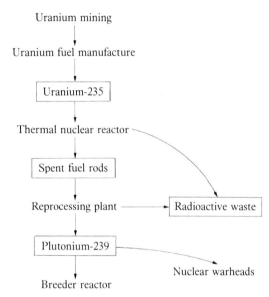

Figure 13.3.1   The nuclear fuel cycle

ACTIVITY

1. The first nuclear reactor in the UK, built in the early 1950s at Calder Hall, did not produce electricity. Why was it needed?

2. Discuss the arguments for and against the reprocessing of uranium fuel rods.

3. Find out how spent fuel rods are transported from nuclear power stations to the reprocessing plant at Sellafield. What safety precautions are taken?

4. Suggest why Dounreay was chosen as the site for the UK's prototype breeder reactor.

5. Explain why water or gas are not suitable as coolants for a breeder reactor.

6. Suggest why it is unlikely that a full scale breeder reactor will be built in the UK in the foreseeable future.

Science, Technology and Society © D. Andrews, 1992

SCIENCE TECHNOLOGY AND SOCIETY

13 NUCLEAR POWER

UNIT 13.4 What can we do with radioactive waste?

Radioactive substances can be dangerous if they are not handled carefully. In Units 4.4 and 13.6 the potential health hazards of radiation are investigated. The amount of radioactive waste produced worldwide is steadily increasing. Most of this waste is produced by nuclear power stations, but smaller amounts come from hospitals, research laboratories and industrial plants. The disposal of this waste will continue to be a controversial issue in the future.

ACTIVITY

Options for radioactive waste disposal

How should we dispose of radioactive waste so that it does not harm us or the environment? The following methods have all been considered. Think about the likely costs and risks involved and list the advantages and disadvantages of each option.

- Load it on to a rocket and launch it into space.
- Surround it in a concrete block and dump it in the ocean.
- Load it into steel drums and bury it 500 m below the land surface.
- Load it into steel drums and bury it 500 m below the sea bed.
- Encase it in concrete and steel blocks and store it at guarded sites on the surface.

TYPES OF RADIOACTIVE WASTE

There are three types of radioactive waste.

- **Low-level waste** is slightly radioactive. It consists of waste gases, liquids and solids such as gloves and containers used in handling radioactive materials. Although the release of low-level radioactive waste is legally permitted below a specified limit, its long-term effects on the environment are uncertain. Radioactive waste passing into the Irish Sea from the Sellafield reprocessing plant is known to accumulate in shellfish.

- **Intermediate-level waste** is more radioactive. It includes material removed from nuclear reactors that have reached the end of their useful life. This must be stored carefully to prevent contamination of the environment.

- **High-level waste** is the most radioactive. It is produced when the uranium fuel rods removed from nuclear reactors are reprocessed. It generates enormous amounts of heat and so must be cooled continuously.

The amounts of intermediate-level and low-level waste in the UK are shown in Figures 13.4.1 and 13.4.2.

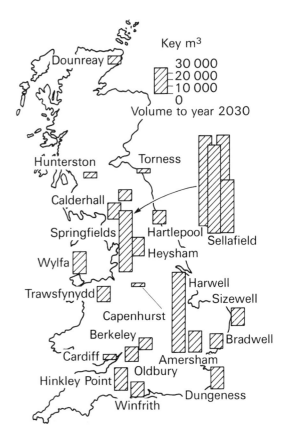

Figure 13.4.1 The estimated volumes of low-level waste in the UK that will have accumulated by 2030.

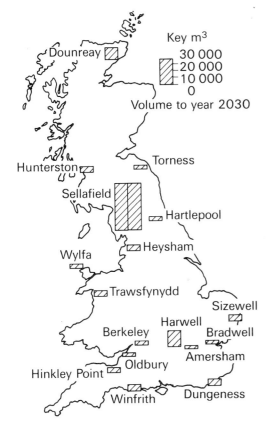

Figure 13.4.2   The estimated volumes of intermediate-level waste in the UK that will have accumulated by 2030.

Table 13.4.1   Radioactive waste production by Radavia's nuclear reactors.

Nuclear plant	Annual production of radioactive waste (m³)		
	High-level	Intermediate-level	Low-level
Alphaville	0	100	900
Betatown	0	60	500
Gammahaven	0	120	1000
Delta	0	150	2000
Electra	0	200	2000
Fanshale	0	50	400
Byfield	40	400	2500

ACTIVITY

Siting a radioactive waste dump – a decision-making exercise

Radavia is an imaginary country with a difficult decision to make: where should its radioactive waste be dumped? At present its six nuclear power stations store their own low-level radioactive waste. This is unsatisfactory and a single dump site is being looked for. Some intermediate-level waste is transported from the nuclear power stations to the reprocessing plant at Byfield, which produces high-level waste. Scientists have recommended that the waste dump should be located deep underground at one of the nuclear reactor sites.

Use the map of Radavia (Figure 13.4.3) to answer these questions.

1. Draw up a table showing the advantages and disadvantages of each nuclear reactor site as a waste dump.

2. Write a short report giving your assessment of each site and include a reasoned argument for the one site you believe is most suitable. You should consider:
   - The geographical features shown on the map of Radavia.
   - The annual radioactive waste production figures for each nuclear reactor (Table 13.4.1).
   - The guidelines for locating a waste dump outlined in Box 1.

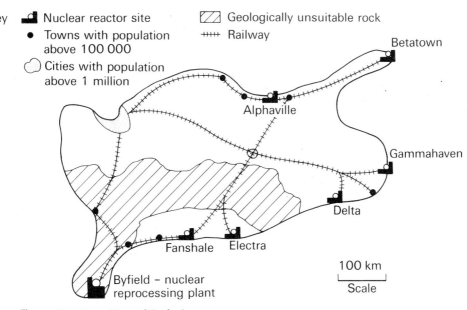

Figure 13.4.3   Map of Radavia.

Science, Technology and Society © D. Andrews, 1992

1. Locating a waste dump

When deciding on the site for a radioactive waste dump, the following factors should be considered.

- **Rock structure** Certain areas of the country are unsuitable because of the type of rock found there. Underground water moving rapidly through the rock strata can increase the risk of radioactive substances contaminating drinking water.
- **Transport links** The site should be easily accessible by rail from all the major nuclear reactors. If the rail journeys for transported waste can be minimised, the costs can be reduced. You may also want to suggest the construction of new rail routes.
- **Public reaction** A common reaction when people hear that a radioactive dump might be sited near to them is 'Not In My Backyard', sometimes shortened to NIMBY. Your choice should take into account the number of people living nearby who might react unfavourably.

2.

'If all the electricity used by one person in a lifetime were generated by nuclear power the resultant highly active wastes could be incorporated into a block of glass this size.' (See Figure 13.4.4.)

UK Atomic Energy Authority

'Between 2500 and 5000 m^3 of intermediate-level waste and between 25 000 and 40 000 m^3 of low-level waste are being produced in Britain each year. Over the past 30 years Britain's nuclear power programme has produced 1570 m^3 of high-level waste.'

United Kingdom Nirex Limited

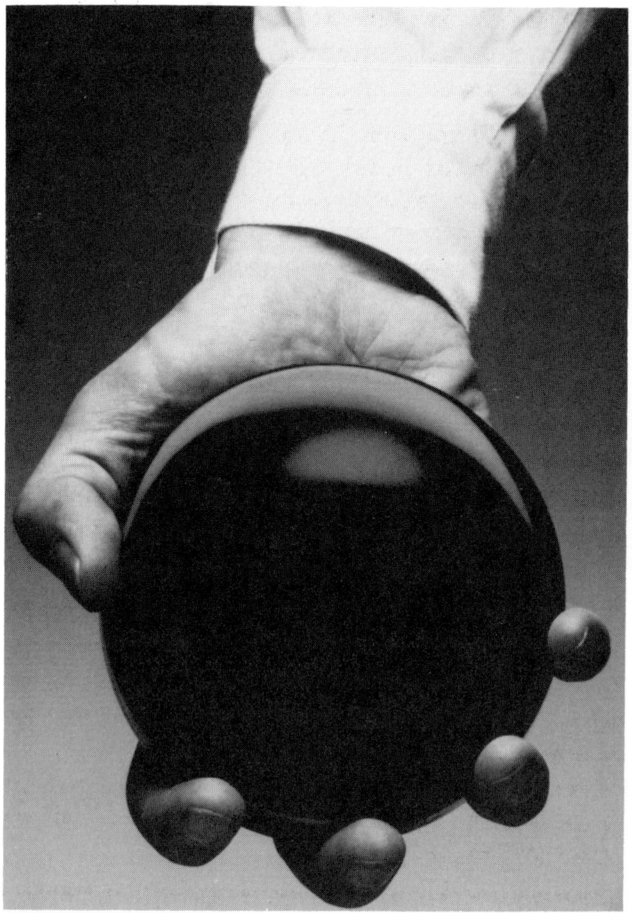

Figure 13.4.4

ACTIVITY

The radioactive waste problem

1 Consider the statements in Box 2, which were made by two different organisations concerning the amount of radioactive waste being produced in the UK. Which organisation's statement minimises the problem? How have they achieved this?

2 Find out about UK Nirex Ltd, the organisation responsible for disposing of radioactive waste. What are their plans for the UK?

3 Friends of the Earth, the environmental pressure group, question the safety of storing radioactive waste in underground dumps. Their view is that it is not possible to predict the chances of radio-isotopes leaking into water supplies because of uncertainties over future movements of ground water. Write off for further information and compare the views of FoE with those of UK Nirex.

13 NUCLEAR POWER

UNIT 13.5 The Chernobyl accident

In the early hours of 26 April 1986 an explosion blew the top off one of the four reactors at the Chernobyl nuclear power station 100 km from the Soviet city of Kiev. A series of fires in the reactor continued until 2 May, releasing radioactive material into the atmosphere. This material was carried on strong winds over a large part of Europe. Thirty-one fire-fighters died from severe burns caused by high doses of beta radiation. In the following weeks 300 local people suffered radiation sickness.

Over 100 000 people in the surrounding area were evacuated. In Kiev iodine tablets were taken as a precaution. The ordinary iodine saturates the thyroid gland and prevents it from absorbing radioactive iodine-131, which would increase the risk of thyroid cancer. Despite these precautions the rate of cancer and birth defects in the area increased dramatically.

The cloud of radioactive material followed an erratic path over western Europe as the wind direction changed (Figure 13.5.1). The amount of radioactive fallout reaching the ground depended on the rainfall at the time the cloud was passing over. In the UK, Cumbria, North Wales, and south-west Scotland were particularly affected because of heavy rainfall. In North Wales, the radioactive isotopes caesium-134 and caesium-137 were deposited on the grass, and consumed by sheep and retained in their bodies. In many sheep the level of radioactivity exceeded government limits, so they were deemed unfit for human consumption. Farmers received compensation from the UK government for these sheep. These restrictions remained in force for over five years after the Chernobyl accident. By 1990 £7 million compensation had been paid.

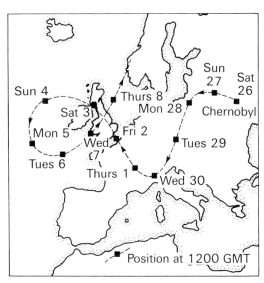

Figure 13.5.1 The approximate path of the radioactive dust cloud released from the Chernobyl nuclear reactor. It passed over the UK in early May, 1986.

'radiation is still pouring into the air from a fire at the plant' (*Daily Mail* 1.5.86)

'radiation... catapulted into the sky and leaking into the immediate area around Chernobyl' (*Daily Mirror* 14.5.86)

'Tests showed that rainwater contained fairly high levels of short lived radiation.' (*Daily Mail* 30.4.86)

'Contamination comes from radiation dust (*Daily Mail* 6.5.86)

'liquid iodine is an antidote against radiation sickness' (*Daily Mirror* 1.5.86)

ACTIVITY

1 Use your understanding of radioactivity (see Unit 13.1) to comment on the impressions given by the extracts in the box that appeared in newspaper reports at the time of the Chernobyl nuclear reactor accident.

2 Has the Chernobyl disaster made future nuclear reactor accidents more or less likely? Give reasons for your view.

3 Scientists' estimates of the number of extra cancer deaths that would be caused by the Chernobyl fallout varied from several hundred up to one million people. Suggest reasons for this very significant disagreement.

4 Suppose that 1000 extra cancer deaths over the next 30 years is an accurate estimate. The estimate for total cancer deaths from all causes over the same period worldwide is 15 million. How does this affect your assessment of the health hazard of the Chernobyl fallout?

5 Find out how the Chernobyl accident happened. In particular, how was responsibility for the accident divided between operator error and faulty design?

13 NUCLEAR POWER

UNIT 13.6 Nuclear power and leukaemia: problems of extrapolation

For over 40 years researchers have made a careful study of the health of survivors of the 1945 atomic blasts at Nagasaki and Hiroshima. These studies have shown that over the following decades the survivors suffered an increased risk of different types of cancer, including leukaemia (see Unit 4.4). Scientists have applied these findings to the problem of the health effects of low-level radiation by assuming a consistent relationship between the dose received and the risk of leukaemia. This relationship can be shown as a line on a graph, which is extended, or **extrapolated**, into regions not covered by the observed data. However there is disagreement among scientists as to what type of line represents the true relationship between radiation dose and the risk of leukaemia. The three different lines shown in Figure 13.6.1 reflect the variation in the assumptions made about the relationship between the variables. A linear relationship assumes that the cancer risk is directly proportional to the radiation dose. A quadratic relationship assumes that increasing the dose by a factor, x, will lead to an increase in the cancer risk by a factor x^2.

LEUKAEMIA CLUSTERS

In some parts of the UK there are unusually high concentrations of leukaemia cases. Some of these clusters occur near nuclear power stations and other nuclear installations such as the Aldermaston research reactor (see Figure 13.6.2). In 1983 James Cutler, a television producer, discovered that the death rate from leukaemia for children under ten in the parish closest to the Sellafield nuclear reprocessing plant was ten times higher than expected. An official report investigating this finding came to no firm conclusion. Can we be sure that the leukaemia cases are a direct result of the radioactive emissions of nuclear reactors? There has been considerable disagreement among scientists over this question.

Some have pointed out that as the number of leukaemia cases is very small the clusters could be the result of chance effects. There are some nuclear power stations that do not have clusters nearby and some clusters are found well away from nuclear reactors. A more recent theory is that low-dose radiation has produced mutations in the gametes of the nuclear plant workers, which increase the risk of leukaemia in their offspring.

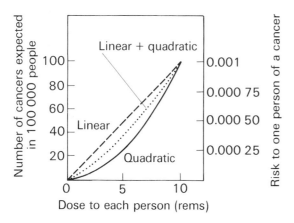

Figure 13.6.1 Three alternative curves for relating the radiation dose a person receives to the risk of suffering cancer.

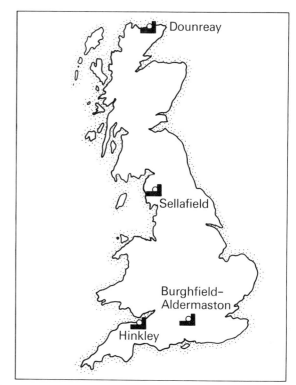

Figure 13.6.2 The four British nuclear sites with leukaemia clusters.

ACTIVITY

Health risks of radiation

1. Plot the data given in Table 13.6.1 on a line graph.

2. What is the relationship between the two variables?

3. Give two differences between the radiation dose received by the Nagasaki and Hiroshima survivors (see Unit 4.3) and the dose received by workers in nuclear power stations.

4. Which of the extrapolated lines in Figure 13.6.1 gives the highest risk for low-dose radiation?

5. How does Figure 13.6.1 help you to appreciate why scientists disagree over the health risks of low-level radiation?

6. Write off for information from British Nuclear Fuels and Friends of the Earth on this issue. Assess the two opposing viewpoints and write a short report on the arguments.

Table 13.6.1 The incidence of leukaemia following exposure to different radiation doses as a result of the Hiroshima and Nagasaki atomic bombs.

Radiation dose (rems)	Incidence of leukaemia (cases per 10 000 per year)
100	1.1
250	2.7
500	5.3
1000	9.9
1500	15.3

13 NUCLEAR POWER

UNIT 13.7 Hot and cold fusion power

At the centre of all stars, including our own Sun, atoms of hydrogen are colliding and fusing together, releasing enormous amounts of energy, which we experience on Earth as light and heat. This is atomic **fusion** and it will only occur at temperatures above 100 million degrees Celsius. Uncontrolled atomic fusion was first demonstrated by scientists in the hydrogen bomb.

Controlling this fusion reaction could provide an exciting new source of energy. Scientists have been investigating the possibility of a fusion reactor over the last 30 years. There would be several advantages of such a reactor over a conventional fission reactor.

- A fusion reactor would be fuelled by the elements deuterium and lithium, which are readily extracted from sea water and the Earth's crust, respectively. Fusion power would be a long-term solution to the depletion of non-renewable fuels.

- Tiny quantities of deuterium can produce enormous amounts of energy. One cubic metre of sea water yields 33 grams of deuterium, enough to produce the energy equivalent of 200 tonnes of oil.

- Unlike fission reactions, fusion reactions do not produce long-lived radioactive waste products. The waste disposal problem is limited to the parts of the reactor that will become radioactive.

- A fusion reactor cannot develop an uncontrollable runaway reaction as occurred in the fission reactor at Chernobyl. If a mistake is made in the running of a fusion reactor, the reactor comes to a grinding halt.

However there are many technical problems to overcome before a fusion reactor becomes a reality. At the very high temperatures required the atoms become stripped of their electrons, forming a plasma. Powerful magnetic fields must be used to contain the plasma and prevent it from making contact with the walls of the reaction vessel, as any contact would rapidly cool the plasma. Experimental fusion reactors called tokamaks are doughnut-shaped vessels surrounded by circular magnetic coils (see Figure 13.7.2). Tokamaks have been successfully used to heat the plasma to the required temperature of 100 million degrees Celsius.

For fusion to occur there are two other requirements. The density of the plasma must reach about 10^{20} particles per cubic metre (1/1000 g per cubic metre), and the period of time that the plasma retains its energy must be between one and two seconds. If these three conditions are met at the same time the plasma will generate more energy than is needed to heat it up. At this point the excess energy, in the form of energetic neutrons, can be transferred to a coolant, which can be used to form steam to drive a turbine.

As well as the technical difficulties, the high cost of developing fusion reactors has to be overcome. An international prototype fusion reactor is to be built in the 1990s but it will cost around £2.8 billion and require the co-operation of many countries.

In November 1991 the experimental fusion reactor known as JET, in Culham in Oxfordshire, achieved a historic breakthrough. A sustained fusion reaction in a deuterium-lithium plasma was maintained for 20 seconds and generated a megawatt of power.

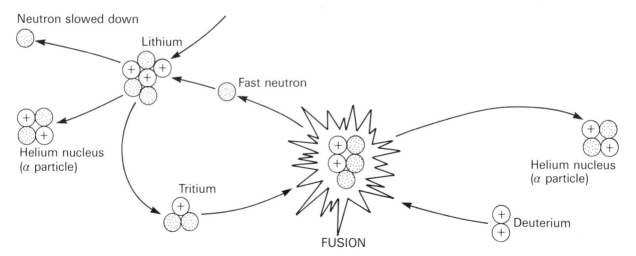

Figure 13.7.1   Deuterium and either lithium or tritium are fed into a fusion reactor. Neutrons produced by the fusion reaction bombard lithium atoms, converting them to titium. The alpha particles are absorbed by the walls of the reactor vessel.

ACTIVITY

Energy from fusion

1. Explain the difference between a fusion reaction and a fission reaction.
2. What three conditions are required for fusion to occur?
3. Suggest why governments are prepared to continue putting money into fusion research despite the difficulty of achieving these three conditions at the same time.
4. What are the advantages to be gained in this type of research by collaboration between several countries?
5. Fusion power has been described as 'an attractive but still distant promise'. Do you agree?

COLD FUSION – A MIRAGE?

In March 1989 two scientists, Stanley Pons and Martin Fleischmann, became world-famous celebrities overnight after they made a startling announcement at a special news conference. They claimed that they had achieved nuclear fusion of deuterium atoms in an electrochemical cell at room temperature and that this reaction produced four times as much energy as it consumed. In the following days cold fusion was described as a remarkable scientific breakthrough that would be capable of solving the world's energy crisis. Scientists immediately attempted to repeat these remarkable results in their own laboratories. Many found that Pons and Fleischmann had made serious errors. One research team concluded that the equipment used by Pons and Fleischmann was not accurate enough to give significant readings. There was also criticism that they had not carried out control experiments using an electrochemical cell containing ordinary water instead of deuterium. The general view was that fusion had not taken place.

The way Pons and Fleischmann used the press to publicise their discovery surprised many scientists, since it is usual for research findings to be submitted initially to a scientific journal. Before it is published the report is scrutinised by a panel of experts to check its accuracy. This process is called peer review and is intended to maintain high standards of scientific research. The press announcement short-circuited this checking process.

ACTIVITY

The media and scientific discoveries

1. What do you suppose were the main reasons for Pons and Fleischman announcing their discovery the way they did?
2. What is the advantage of the normal practice of scientists attempting to replicate each other's results?
3. Suggest why this scientific announcement made such an impact on the media.
4. Carry out a survey comparing the way an event with scientific implications, a particular scientific advance, or a novel technological application is reported on TV, in a quality newspaper, in a tabloid newspaper and in the weekly magazine, *New Scientist*.

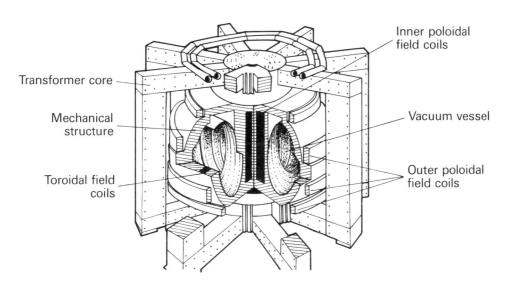

Figure 13.7.2   Schematic diagram of the Tokamak.

Teacher's notes

The following guidance notes are intended to anticipate teachers' concerns, recognising that in some cases the topic covered may be unfamiliar. Answers are provided to numerical questions and to the more factual questions where the answers are not readily accessible from the text. Where appropriate, guidance is provided on questions designed to elicit the students' views. Further reading and sources of additional information are also given. The addresses of the organisations referred to are given in 'Sources of further information' on page 181.

1 SCIENCE – THE KEY TO THE UNIVERSE?

Unit 1.1 What is science?

The purpose of this unit is to provide a basic understanding of the scientific approach and to introduce the idea that science develops through the construction and replacement of paradigms.

The Structure of Scientific Revolutions by T S Kuhn (University of Chicago Press, 2nd edition, 1970) has been very influential in this field and is recommended for the interested teacher.

Unit 1.2 Religion and evolution

ACTIVITY: Creationism

It is important to emphasise how the new idea of evolution caused a revolution in the scientific status of biology. It changed from a purely descriptive study to an analytical science, with the necessary concepts to give a coherent and comprehensive explanation for the differences in the forms and functions of living organisms.

ACTIVITY: The great debate

1 The term 'obscurantist' means someone who tries to prevent reform or enlightenment. Here it refers to a tendency to oppose scientific enquiry and explanation.

2 LAUNCHING TECHNOLOGY

Unit 2.1 The space race: competition or co-operation?

ACTIVITY: Space missions – economic and political influences

1 The lower cost and the greater military and scientific value of manned space stations made them more attractive to the politicians.

2 There was greater technical sophistication and public prestige attached to a manned mission to the Moon.

4 The main advantages are reduced costs, sharing of technological expertise and the development of warmer political relations between the countries involved. Because of the military importance of space technology there may be a reluctance to discuss technology that could reveal the country's capabilities.

Unit 2.2 The *Challenger* Shuttle disaster

ACTIVITY: Shuttle problems

1 The American public were anxious about the Shuttle's cost to the US budget.

2 To increase the public's identification with the programme and reduce concern about NASA's spending.

3 a) Freezing temperatures make the rubber brittle. It is more likely to crack under the pressure during take off.

b) Placing a rubber band in a beaker of iced water will demonstrate this effect. This demonstration was originally used by a member of the Shuttle investigation committee, Richard Feynman, at a press conference.

c) NASA flight directors acted on an over optimistic estimate of the chances of a critical accident, an unquestioning faith in the reliability of the Shuttle's design and the political pressure to maintain a target of a minimum number of launches.

ACTIVITY: To launch or not to launch – a role play

A warm-up exercise is sometimes useful to reduce students' anxiety about a role-play exercise. This could involve the teacher and an extrovert student volunteer conducting a short role play based on a meeting between two of the characters.

Unit. 2.3 Future space exploration: a mission to Mars?

The information given for the various characters in this simulation is based on facts about real people. The comments made, the incidents reported, and the research findings are factually accurate. The article in the February 1991 edition of *New Scientist* gives further details. See notes on role plays for Unit 2.2.

Unit 2.4 Industry and innovation

ACTIVITY: A national fibre optics network

1. • redolent – suggestive of
   • omnipotent – all powerful
   • compatibility – ability to be used in combination with

4. • teleshopping – time-saving and more convenient than making a trip to the shops but it may lack the sense of 'occasion' of a shopping trip.
   • teleworking – avoids the hassles of commuting and offers the possibility of work for people who are housebound e.g. mothers of young children, or the disabled. Lack of direct social face-to-face contact with work colleagues is a drawback.
   • teleconferencing – relatively inexpensive way of arranging a discussion between people who are thousands of kilometres apart. You would only be able to see the person who is talking at the time.
   • electronic freight transport – inexpensive but can only be used for products consisting of written or pictorial material.

Unit 2.5 Industry and the environment: bauxite mining

ACTIVITY: Sectors of industry

Privately owned	Publicly owned
National Power	Nuclear Electric
Harrods	National Health Service
British Telecom	British Coal
	The Post Office
	British Rail

The swimming pool is probably publicly owned. The students' school is either publicly or privately owned.

Primary sector British Coal
Secondary sector Nuclear Electric, National Power
Tertiary sector British Rail, school, swimming pool, the Post Office, National Health Service, British Telecom, Harrods

ACTIVITY: A bauxite mine for Saint Francis?

1. The teenager and the owner of a small store might be expected to welcome the increase in economic activity and the availability of employment. However, the possibility of enthusiastic support for the cause of environmental protection could not be ruled out.

2. See the notes on role play for Unit 2.2, page 162.

3 INFORMATION TECHNOLOGY

Unit 3.1 Trends in information technology

ACTIVITY: Communication technology

The order of invention: face-to-face speech; visual signalling – smoke signals; letters transported by hand; letters transported by horseback; visual signalling – flags, flashing lamps; letters transported by train, wireless telegraphy using morse code; telephone; letters transported by air; radio; telex; television; electronic mail from one computer terminal to another; fax.

Unit 3.2 From valves to transistors to microchips

ACTIVITY: Social impact of advances in communications

4. The line of the graph will flatten out as the limit imposed by the principles of physics is reached, where electrons leak out of the microcircuits.

Unit 3.3 Computer applications

ACTIVITY: Using computers

1. a) **Newspaper and book publishing** Computer typesetting and DTP allow for improved efficiency. The transition from traditional hot metal typesetting was opposed by printworkers' industrial action.
   b) **Supermarkets** Optical reading of bar codes has improved monitoring of sales and stock control, as well as speeding up the check-out queues.
   c) **Banks and building societies** Electronic fund transfer at the point of sale (EFTPOS), e.g. SWITCH and CONNECT, and automatic cash dispensers are reducing the need for written cheques.
   d) **Education** Students' views on the value of computer-assisted learning as compared to traditional teaching methods should allow for lively debate and may raise the question, what can a teacher provide that a computer cannot?

Unit 3.4 Artificial Intelligence

The Creative Computer, by Donald Mitchie and Rory Johnston (Pelican, 1985) provides a more extensive exploration of the issues raised in this unit.

ACTIVITY: The potential of AI

4. The experience from Glasgow is that patients enjoy sitting at a terminal and answering questions, and that the doctors value the computer's statistically validated probabilities for various

possible diagnoses e.g. ulcer, cancer. Some may see this as handing over too much responsibility to the computer.

Unit 3.5 The Data Protection Act: who is holding information about you?

ACTIVITY: **Computer security**

1 Banks, educational institutions, local authorities and central government (at least).

2 Through the completion of forms required by organisations or the disclosure of data by one organisation to another.

3 Receiving an unsolicited, computer-addressed personal mailshot. Mailshots are usually a sure sign that an address database has been used.

4 Speed of access and transmission, ease of amendment, reduced bulk of stored records.

The address of the Data Protection Registrar is given on page 181.

Unit 3.6 Could a computer do your job?

ACTIVITY: **Cutting jobs in the salaries department**

1 Jobs created by growth of IT include: programmers, systems analysts, computer operators and computer engineers.

2 There are arguably very few jobs that have not benefited directly or indirectly from the assistance of IT.

7 The possibility of errors due to faulty software could be a counter-argument to the charge of human clerical mistakes.

8 Inevitably, if people have suffered prolonged unemployment because their job has been replaced by computer technology, they are likely to feel the same hostility and suspicion as shown by Mr Ramsey.

ACTIVITY: **Computers and jobs**

An alternative use of this Activity is for the students to write their own short play, inspired by the scenes suggested.

Unit 3.7 Marketing innovations: satellite television

This unit can be used to illustrate the concepts of 'technology push' and 'market pull' as forces leading to the introduction of innovations. The technology push gathered momentum as the costs of buying a dish and setting up a satellite television station fell. The market pull resulted partly from the sense of personal prestige that the consumer gains from the possession of publicly visible high technology equipment and partly from the desire to have a wider choice of viewing.

4 WAR, TECHNOLOGY AND SOCIETY

Unit 4.1 Warfare and technological development

ACTIVITY: **High-tech war?**

2 • 245-T (Agent Orange), a herbicide defoliant, and napalm, a chemical incendiary agent, were both used by the USA against the Vietcong in the Vietnam War. This brought much condemnation of the US strategy, leading to a strengthening of the public opposition to the war in the USA.

• Spy satellites were used to locate missile launchers in the Gulf War, 1990.

• Computers for code-breaking and radar were first used by the Allies in World War II.

• Tanks used in World War I were unreliable and not a significant factor in the defeat of Germany.

• The USA's 'smart' bombs used in the 1990 Gulf War allowed precision bombing.

Unit 4.2 The atomic bomb: from Einstein to Hiroshima

Brighter than a Thousand Suns, by A Jungk (Penguin, 1960) would be suitable reading for motivated students. It is a fascinating and detailed account of the development of atomic weapons from 1910 to 1950.

ACTIVITY: **Secrecy and science**

1 Sharing information and collaboration on developing ideas helps to increase the rate of scientific progress.

2 The Official Secrets Act could slow down the rate of scientific progress.

3 Fuchs' sympathies with the political aims of communism were more important to him than allegiance to Britain.

Unit 4.3 Atomic bomb decisions

ACTIVITY: **How should the atomic bomb be used?**

1 1, 2 and 3 would have been used to justify the military use of the bomb while 4 and 5 were strong opposing arguments.

4 The voting was as follows:

 1 15%
 2 46%
 3 26%
 4 11%
 5 2%

Unit 4.4 The medical effects of nuclear war

ACTIVITY: Medical impact of radioactivity on pregnancy

1 a) Group C   1 b) Group A

2 a) A 10%   B 1.5%   C 0.9% 1·8%
   b) A 43.3%   B 8.8%   C 6.2%

3 The group with highest exposure (A) suffers from the highest pre- and post-natal mortality rates, owing to the fatal effect of radioactivity on dividing cells.

4 Calculating the percentage of each sample suffering from retardation indicates that the hypothesis is supported.

ACTIVITY: Radioactivity and growth

1 a) reduced height
   b) no significant impact on height
   c) no significant impact on height.

2 The radiation has a more serious impact on the under-five age group because cell division is more rapid at this stage and the cells are more vulnerable to damage.

Unit 4.5 Controlling the arms race

This unit should be supplemented with topical information concerning issues on defence and arms control, for example, the impact of the peace dividend on the arms industry.

ACTIVITY: The arms control controversy

1 The combination of the necessary scientific expertise and the desire for status as a world power. In the case of less powerful countries, e.g. India, the aim was to gain military superiority over neighbouring enemy powers.

3 a) With some countries unwilling to sign the NPT, its usefulness can be questioned.
   b) Proliferation increases the possibility of use, since nuclear weapons could be used in a larger number of conflicts.

4 Satellite photographs can identify missile launchers.

Unit 4.6 Chemical weapons

ACTIVITY: Political, military and technological issues

2 The need for chemical weapons is avoided if a country has nuclear arms.

5 An unannounced visit to the plant and an inspection to observe and analyse the raw materials and the products.

8 There will always be loopholes in any monitoring of chemical stocks, especially if it is a voluntary code.

10 The attitudes of scientists working in this field are similar to those of the physicists involved in the Manhattan Project to develop the atomic bomb.

---

5 THE REPRODUCTION REVOLUTION

Unit 5.1 Natural selection at work

ACTIVITY: Natural selection at work

1 The regions in which malaria is endemic are also those that have a high frequency of the sickle cell gene. This is evidence for the protective value of the gene in these areas. In other regions the sickle cell gene does not provide this advantage and so selection operates to remove it.

2 If a person has sickle cells they are less likely to suffer from malaria (probability = 0.28) than someone who lacks sickle cells (probability = 0.46).

Unit 5.2 Basic genetics

ACTIVITY: Breeding by coin tossing

2 The ratio of 3 : 1 should be more closely approximated with the pooled results.

3 Larger numbers ensure that random variation is minimised.

4 The coin represents the paired genes on the chromosomes before they are separated by meiosis. Coin tossing represents the equal probabilities of the two types of gametes being formed.

Unit 5.3 Harmful genes 1: sickle cell anaemia

The question of harmful genes needs to be handled carefully, especially if a student comes from a family in which the known risk of an inherited disorder is high. Some students may have already had to come to terms with the knowledge that they are carriers.

Unit 5.4 Harmful genes 2: cystic fibrosis

See notes to Unit 5.3.

ACTIVITY: Choices for CF carriers

6 Students should be aware of the cost-effectiveness argument that some politicians and health economists have espoused. They argue that it is

Science, Technology and Society © D. Andrews, 1992

much cheaper to use screening programmes to 'weed out' the genetically unhealthy before they are conceived than to treat them after they are born.

7 Presumably there will be many willing volunteers for risky treatments given the present prognosis for CF sufferers.

Unit 5.5 Amniocentesis and Down's Syndrome

ACTIVITY: Your views on amniocentesis

1 The graph shows that the risk of trisomy 21 begins to increase significantly for mothers aged over 35.

2 CVS allows earlier detection of chromosome abnormalities and therefore an earlier and perhaps less distressing abortion.

5 Parents who might opt for abortion because of the anticipated financial and emotional burdens of caring for a handicapped child might be less likely to do so if support could be guaranteed.

6 By assuming that the handicapped child would become an unnecessary financial burden on society, these doctors believe that the cost of detection can only be justified if abortion of an abnormal fetus automatically follows.

10 It was realised that the genetic relationship with the Mongol race was unfounded. The term might also have unfortunate racist overtones.

ACTIVITY: Role play

See the notes for Unit 2.2 on page 162.

Unit 5.6 Images of disability

ACTIVITY: Behind the myth

2 Discrimination and low expectations can restrict opportunities for the disabled.

3 The title *The Dungeon of Love* implies that the disabled can be smothered and imprisoned by the pity and sympathy that they receive from other people. This locks them away from contact with the rest of the population and prevents them from learning independence. Carers may have a tendency towards overprotectiveness.

Unit 5.7 Mass screening for inherited disorders

ACTIVITY: Your views on mass screening

1 The assumption is made that a person who knows that they are a carrier will avoid having a child with a partner who is also a known carrier.

2 This depends on the view taken on the rights of the state to take public health measures that interfere with individual freedom. There is a similar argument concerning screening tests for other diseases, e.g. cancer.

3, 4 Voluntary testing is more acceptable but will not detect all the carriers in the population, so would not be as effective in reducing the incidence of inherited disorders.

5 Ethnic groups at risk are likely to feel stigmatised. Other children's prejudiced views towards these groups may be reinforced.

7 b) Break off the engagement, agree to a marriage that will be childless, or use prenatal testing and selective abortion to avoid the birth of a child with the disease.

9 a) The possibility of individuals having 'genetic identity' cards showing the recessive harmful genes that they are carrying has been suggested. If the government has access to this personal data it might encourage moves to introduce a 'license to reproduce', especially if it would lead to savings in the health service budget.

b) Public awareness, coupled with the necessary legislation concerning the privacy of personal genetic data, should be effective in preventing this trend.

Unit 5.8 Screening embryos and fetuses

ACTIVITY: Your views on the testing controversy

1 The doctors assumed that there would be no objections to gathering a more detailed picture about the health of the fetus.

3 The justification is that the parents will have to shoulder the burden of caring for a child that is likely to have a short and possibly painful life.

ACTIVITY: Your views on screening embryos

1 There would be cost savings to the health service from reducing the numbers needing treatment for these conditions.

2 These illnesses are a result of environmental influences as well as an inherited susceptibility. Genetic screening would be ineffective in reducing their incidence if environmental causes were predominant.

3 Complete parental freedom over the choice of their child's characteristic has been presented as a fundamental right by some commentators. Others see it as treating a child as a commodity.

Unit 5.9 Genetic counselling

ACTIVITY: **Personal reactions**

1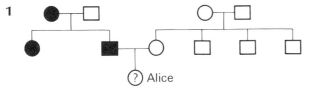

2 The disease appears in all three generations.

3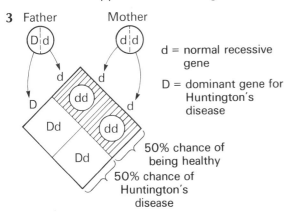

6 If the person shows suicidal or depressive tendencies the doctor may believe that the person could not cope with the knowledge that he/she is suffering from a terminal illness.

---

6 IMPROVING THE SPECIES?

A deeper analysis of the legal issues touched on in this section is found in *Medicine, Patients and the Law*, by Margaret Brazier (Penguin, 1987).

Unit 6.1 Test tube babies

ACTIVITY: **Opposition to IVF**

2 Sir John Peel was implying that a child conceived through IVF would suffer from psychological distress when it realised the artificial nature of the techniques used.

3 Supporters of IVF argue that the acute distress that can be experienced by an infertile couple requires a medical treatment if it is available.

Unit 6.2 Embryo research choices: Germany and the UK

ACTIVITY: **Embryo research – the political debate**

1 A Bill is the proposed new legislation, which becomes an Act after Parliament has amended it and a majority of MPs have voted for it.

2 Normally the MPs have to vote according to the instructions of the Party Whips. This issue was seen to be a matter of individual moral judgement by MPs rather than an official party political policy.

6 The German view on any attempts to interfere with the gene pool is influenced by its association with the theories of racial superiority and the elimination of 'substandard' members of the human species proposed by the Nazis. The German attitude can be seen as reaction to this. The British do not share this historical background.

Unit 6.3 Freezing embryos

ACTIVITY: **Your views on frozen embryos**

1 There is no clear evidence of the viability of human embryos stored over long periods. If an embryo is stored over decades the original mother may be past childbearing age.

2 The options are either donation of the embryo to another couple, or disposal.

3 The ethical objections focus on the problems of disposal of frozen embryos and inherent repugnance at removing and storing a potential child.

5 The child would experience the same dilemmas as if it had been adopted. The child's feelings would be further influenced by the knowledge that its prenatal development took place in its adoptive mother's uterus. Anonymity of the donor parents might avoid some of the difficulties.

Unit 6.4 Surrogate motherhood: rent a womb?

ACTIVITY: **Your views on surrogacy**

3 The strongest claim would occur in the case of the surrogate mother who is also the ovum donor and the child's biological mother.

4 The child may claim inheritance from the biological mother (2) or parents (3). In case 1 there would be no legal claim by the surrogate mother.

8 For example, a woman may desire to avoid the health risks and/or the time off from her career resulting from pregnancy. It has been suggested that this could lead to a wealthy social stratum delegating their child bearing in a similar way to the historical use of 'wet nurses' for breastfeeding.

Unit 6.5 Genetic engineering

ACTIVITY: **Controlling the release of GMOs**

1 GMOs would not be toxic.

2 Competition with chemical pesticides might be judged to be too strong.

Science, Technology and Society © D. Andrews, 1992

3 Uncertainty over the ecological impact of a novel species on other organisms.

Unit 6.6 Gene therapy

It is hoped that students will gain a picture of the useful medical applications of this technique as well as the potential for its abuse.

ACTIVITY: Gene therapy out of control?

As somatic cell therapy, several of these examples are equivalent to life-long drug treatment. Germ line gene therapy would clearly have a more fundamental and worrying impact on the nature of the human gene pool.

Unit 6.7 Who owns the human genome?

ACTIVITY: Copyrighting the human genome

Further discussion of the Human Genome Project can be found in the *New Scientist*, 11 August 1990.

1 1500 scientists.

2 Either from governments or from private investment.

3 The author's work can only be reproduced with permission and on payment of a fee.

4 a) Both have used their energies and creativity in the enterprise, which has produced something that others might wish to use.

 b) The scientist has discovered knowledge that others might have uncovered, whereas the author has produced a unique product that could not be replicated without copying.

7 FEED THE WORLD

Unit 7.1 Good for you?

ACTIVITY: Achieving a balanced diet

1 **Protein** for growth and repair of cells; **carbohydrate** for energy; **fat** for energy and as a structural component in cell membranes, **fibre** to protect against intestinal illness, **vitamins** and **minerals** for a variety of roles in cell metabolism.

2 a) Excess food is stored as fat and the person gains weight.

 b) The shortfall in energy needs is made up by metabolising stored fat, resulting in weight loss.

3 The total energy value (in kilocalories) of each portion is given below.

jam doughnut	300
Mars® bar	295
slice of bread and butter	200
can of fizzy drink (not sugar-free)	180
small packet of crisps	145
fried egg	130
small pot of strawberry yogurt	120
three fish fingers	150
digestive biscuit	70
medium-sized banana	65
half a honeydew melon	60
medium-sized apple	50
three boiled carrots	15

4, 5 The discrepancy between the two rank orders will highlight the fact that the most 'attractive' foods are often the most fattening. This could usefully lead into a discussion on the types of foods (if any) that the students regard as both attractive and healthy.

6 *Diet* published by Longman is a suitable program.

7 The apple clearly has other benefits, such as a greater fibre content and no fat.

8 The Health Promotion Unit of your Area Health Authority should provide free information on healthy nutrition and other health-related topics.

Unit 7.2 Food additives

ACTIVITY: Differing views on food additives

Further information can be found in the many books that have been published on food additives, such as *E for Additives*, by Maurice Hanssen.

2 Beta carotene is the natural additive that has replaced tartrazine.

3 The students should realise the importance of a control group. This group is told that they will be on a tartrazine-free diet but in fact are not, to exclude the possibility of a placebo effect causing improved behaviour. The experimenters judging the children's behaviour change should not be told whether the children are in the control or the experimental group. This avoids the experimenters' expectations influencing their observations of behaviour. This is a double-blind experimental design (see also Unit 10.2, Testing drugs on humans and animals).

There has been some serious doubt over the validity of hyperactivity as a medical syndrome. *The Myth of the Hyperactive Child* by P Schrag and D Divoky, (Penguin, 1979) considers the American controversy over the use of amphetamines to treat this condition. A well-motivated student would gain from reading it.

5 Babies' lower body weights make them more sensitive to the concentration of BHT (or other additives) in foods, since the ratio of BHT to body weight is higher.

Unit 7.3 Food advertising and labelling

ACTIVITY: Match the label

1 To allow consumers to make a complaint.

2 (a) packet soup (b) Branston pickle (c) cereal bar

ACTIVITY: **TV food and drink advertising survey**

This approach can be applied to surveys of food advertising in other media, e.g. newspapers, radio, posters.

Unit 7.4 Mycoprotein

ACTIVITY: **Marketing mycoprotein**

1 The higher fibre content and lower fat content make mycoprotein healthier than beef.

6 Animals were used to test toxicity before human trials began, for ethical reasons. The risk of serious side effects was unknown and therefore it was felt to be wrong to use humans first. Further information about mycoprotein can be obtained from Marlow Foods Ltd (see 'Sources of further information').

Unit 7.5 Intensive agriculture and the environment

The Soil Association will provide information on organic farming (see 'Sources of further information').

ACTIVITY: **Coping with pests**

3 The change in the crop species planted each year prevents populations of specific pests from becoming established.

ACTIVITY: **Biological applications**

1 Rare gene mutations that confer resistance on the insect will survive when the population is exposed to the pesticide. Non-resistant insects will be weeded out and the resistant insects will form the next generation.

Unit 7.6 The Green Revolution

Further reading:
Our Common Future, the Report of the World Commission on Environment and Development, (Oxford University Press, 1987)
The Third World Tomorrow, by Paul Harrison (Pelican, 1983)
The Gaia Atlas of Planet Management, edited by Norman Myers (Pan, 1985)

ACTIVITY

1 The principle of consultation with the community and the selection of low-tech solutions, for example, draught buffalo instead of tractors, would be suitable strategies.

3 The comparison between Africa and Western Europe shows that an increase in fertiliser use does not necessarily lead to a corresponding increase in food production.

8 MEDICINE, HEALTH AND SOCIETY

Unit 8.1 Health care: curative and preventative approaches

Further reading: *The Sociology of Health and Medicine*, Nicky Hart (Causeway Press, 1985)

ACTIVITY: **Changing patterns of disease**

3 The major trends are:

- decrease in the proportion of deaths caused by infectious disease. This is due to improved nutrition (increased intake of protein, vitamins and minerals), better housing conditions and the introduction of antibiotics and vaccinations.
- the increase in the proportion of deaths due to neoplasms (cancer) and heart disease, the so-called diseases of affluence.

Unit 8.2 Appropriate health care: barefoot doctors

ACTIVITY: **Victor Charca**

2 They might be concerned about their inability to make accurate diagnoses of serious illnesses, which could lead to the death of the patient.

3 Regular training courses to update their knowledge and a comprehensive manual to assist in their work are invaluable.

4 Major benefits include local knowledge of the community and a greater understanding of the needs and fears of individuals.

5 This opens up the wider question of the relationship between health and politics. In particular, the students should consider the argument that a more equitable distribution of land is a prerequisite for improved health in many developing countries.

Unit 8.3 AIDS/HIV campaigns

The main aim of this unit is to demonstrate that poverty, inadequate health services and illiteracy are major problems in the fight against AIDS in developing countries. In contrast to Europe and North America, the heterosexual populations of developing countries have already experienced a high incidence of HIV infection, illustrating that vaginal intercourse is as likely a mode of transmission as anal intercourse.

The wider issues, such as reducing the risk of infection by means of safer sex practices, deciding whether to have an HIV antibody test, and the stigma attached to HIV positive individuals, may be touched on during discussion. Leaflets produced by the Terrence Higgins Trust and the Health Education Authority should provide background information on these issues.

Science, Technology and Society © D. Andrews, 1992

ACTIVITY

1 a) No b) Yes c) No d) No e) No f) Yes g) Yes
   h) Yes, but only if fresh blood is transferred, blood to blood.

4 a) Leaflets, distributed to literate community leaders; oral presentations with visual aids; posters with bold slogans.
   b) TV and other media advertising; health education in schools and wider community.

Unit 8.4   Health and social class

ACTIVITY: Patterns of disease

1 a) **Lung cancer**, **bronchitis** and **emphysema** – smoking and air pollution.
   Diabetes – higher levels of obesity, high sugar diet and stress.
   Pneumonia – poor nutrition and housing conditions reducing resistance to infection.
   Peptic ulcer – stress of inner city life, poor diet.
   Accidents – greater hazards on the busier roads and in less safe housing.
   Cirrhosis of the liver – excessive alcohol intake. Alcoholics may find it hard to get work and accommodation and thus move into the cheaper inner city areas.
   b) Heart disease is lower than the natural average. It is often given as the cause of death, which could otherwise be described as a 'natural' death from old age. In the inner city there is a higher proportion who die earlier of other specified illnesses.

2 Bromley has a much lower death rate overall compared to England and Wales as a whole. Bromley is relatively affluent, which correlates with less smoking and better living conditions. Air pollution on the outskirts of London is less than in inner London.

3 The higher incidence of lung disease suggests that air pollution is more serious in East London.

4 55%. This is a much higher proportion of the total deaths compared to 100 years ago, since the serious infectious diseases that were the major killers then are now preventable (see Unit 8.1).

ACTIVITY: Infant mortality and occupational class

1 a) Infant mortality increases from class 1 to class 5.
   b) Differences in social conditions include poorer living conditions and nutrition.

2 a) A below-average birth weight can result from the pregnant mother's poor nutrition and cigarette smoking, and lack of antenatal monitoring and advice.
   b) Birthweight falls from class 1 to class 5.

   c) This trend could be attributed to the poorer diet and higher levels of smoking in class 5.

4 An early visit to the antenatal clinic can detect potential problems and provide important advice on nutrition and life-style.

5 An early first visit to the antenatal clinic is more common among the higher occupational classes. This link could be due to greater anxiety about the pregnancy, or a life-style and educational background that promotes more confidence in making contact with the health services.

6 Improved access to antenatal care allows better monitoring and advice, thus reducing complications during pregnancy and labour and the likelihood of infant ill-health. Graphs showing the correlation between the various factors would be the most informative.

Unit 8.5   Attitudes to abortion

This unit will raise many sensitive issues and will probably reveal a wide diversity of opinion. Further research would give students an exposure to the strength of public opinion. In discussion, the need to respect others' opinions needs to be stressed to students. Pressure groups such as the Brook Advisory Centre (pro-abortion) and the Society for the Protection of Unborn Children (anti-abortion) will provide leaflets to provoke further debate.

Unit 8.6   Medical use of fetal tissue

The Parkinson's Disease Society provides updates on the progress of new treatments, including fetal tissue implantation (see 'Sources of further information'). *Awakenings*, by Oliver Sacks (Picador, 1986), uses a number of fascinating patient case studies to give a vivid description of the difficulties of using the drug L-dopa to treat Parkinson-type symptoms. In 1991 a film of the same name, based on the book, was released.

ACTIVITY: Your views on fetal tissue implants

1 The term 'implant' gives an impression of cells being inserted into an existing organ whereas, in most people's minds, 'transplant' implies the removal of an organ and its replacement by a donor organ. 'Implant' gives a more accurate picture of the operation and reduces the likelihood of confusion with the nonsensical notion of a 'brain transplant'.

11 There is a possibility that the woman would be more likely to give consent if the tissue were to be used for a medical treatment rather than research. However the reverse may happen.

12 The separation of the two operations ensures that no undue pressure is brought to bear on the woman seeking an abortion.

9 WORLD POPULATION – PROBLEMS AND SOLUTIONS

Unit 9.1 Population growth: the demographic transition

ACTIVITY: **Demographic transition**

2 c) Stage 1: up to 1860
    Stage 2: 1860–1880
    Stage 3: 1880–1940
    Stage 4: 1940–present (interrupted by a 'baby boom' 1945–1970).

3 UK: stage 4; Japan: stage 4; Chile: stage 3.

4 The demographic transition occurred earlier and more slowly in the UK. This was due to the gradual improvement in economic and health conditions. In Japan, the medical technology and public hygiene measures that had been developed already in the West could be implemented more rapidly. Slow economic growth has delayed the demographic transition in Chile.

Unit 9.2 Population pyramids

ACTIVITY: **Age structures in developed and developing countries**

2 a) A 98 million, B 103 million
   b) A 30 million, B 58 million
   c) A 3.5 million, B 12.5 million
   d) A 8.5 : 1 B 4.6 : 1

3 a) B
   b) It will decline.
   c) The problems of financing pensions and care of the elderly will increase.

4 A 1 : 1 B 3 : 1

5 A The children will be a drain on the country's economy. There is less opportunity to finance education and there will be pressure for the use of child labour.

ACTIVITY: **The UK's changing age structure**

2 Improved life expectancy due to improved living conditions, public hygiene, nutrition and medical advances.

3 The excess of females over males in the higher age groups reflects their longer life expectancy. This may be due to genetic, environmental or hormonal factors providing protection against the ageing process.

4 The size of the 16–24 age group in 1961 was relatively small. This reflects a reduced birth rate during the war. The under-16 age groups show the 'baby boom' of increased post-war prosperity.

Unit 9.3 Contraceptive choice

Additional information can be obtained from the Health Education Unit of your local Health Authority or from the Family Planning Association (see 'Sources of further information'). For further reading, *Contraception: your questions answered*, by John Guillebaud (Churchill Livingstone, 1986) is recommended.

ACTIVITY: **The impact of contraceptives**

1 Condoms do not have to be supplied by medical workers.

3 Concern about AIDS has made some people switch to using condoms; the publicised health risks of the pill have made some women reluctant to use it.

4 a) IUD. Advantage: once fitted it does not require further attention, apart from regular checks. Disadvantage: the need for medical supervision in fitting and checking.
   b) Sterilisation. Advantage: Relatively cheap, with medical intervention necessary only once. Disadvantages: Psychological factors and cultural resistance.
   c) Pill. Advantage: Reversible and convenient. Disadvantage: Possible health risks.

ACTIVITY: **Suitable contraceptive methods**

1 Unsuitable: sterilisation (in the short term), because they are both too young.
   Suitable: any other method depending on personal preferences.

2 Unsuitable: the pill is not advisable because of her age and smoking habits. The IUD is not suitable because she is childless.
   Suitable: any other method, including sterilisation if she wishes not to have children.

3 Unsuitable: any non-permanent method.
   Suitable: IUD or sterilisation.

4 Suitable: the condom. He may be HIV positive and this method is the only one that could protect his partner from infection.

5 Unsuitable: the IUD and sterilisation, as she is childless. The diaphragm and condom would be unacceptable as they interfere directly with lovemaking.
   Suitable: the pill does not directly interfere with lovemaking.

6 Unsuitable: the IUD can cause pelvic inflammatory disease; the pill is not suitable for women with high blood pressure.
   Suitable: condom or diaphragm; sterilisation if she does not want more children.

7 Unsuitable: the IUD gave her cramps; the pill is 'artificial chemicals'; sterilisation – she may want more children.
   Suitable: condom or diaphragm.

Unit 9.4 Population control policy

The pressure groups Population Concern, and the International Planned Parenthood Federation, can provide information on current family planning campaigns from around the world (see 'Sources of further information').

This activity should make students aware of the cultural and economic constraints on implementing a family planning programme.

---

10 A PILL FOR EVERY ILL?

Unit 10.1 Drug development: the sulphonamide story

ACTIVITY: Sulphonamide

1. a) 'streptococcus cultures' – streptococcal bacteria grown in the laboratory
   b) 'chemotherapeutic effects' – effects of using chemical treatments
   c) 'vivisection' – performing experiments that involve dissection on living animals
   d) 'therapeutic results' – results of the treatment

2. 'In vivo' experiments are on living animals. 'In vitro' experiments take place in a test tube.

3. There is no guarantee that a chemical that is effective in killing bacteria in the test tube will be as effective in the body of a living organism. Unless animal tests are carried out the toxicity of the chemical cannot be assessed.

Unit 10.2 Testing drugs on humans and animals

ACTIVITY: Your views on drug testing on animals

The Association of the British Pharmaceutical Industry, the RSPCA and the National Anti-Vivisection Society will provide relevant literature to extend the debate (see 'Sources of further information').

ACTIVITY: Your views on human drug trials

1. a) If the groups are not closely matched, a true comparison cannot be made and the results are invalid.
   b) This ensures that the patients' expectations do not influence their perception of the drug's effects.
   c) This ensures that the doctors' expectations do not influence the way they record the patients' progress.

4. The type and frequency of the side effect, dosage of the drug, relevant patient characteristics, e.g. severity of disease, other drugs used.

5. A voluntary scheme inevitably misses a high proportion of adverse drug effects.

Unit 10.3 Tackling the drug problem

For further information, TACADE (Teacher Advisory Council for Drug Education) or your local Health Education Authority can provide leaflets (see 'Sources of further information').

Unit 10.4 Drugs in the developing world: inappropriate technology?

Further reading: *Pills, Policies and Profits*, by Francis Holt (War on Want, 1985) and *Bitter Pills*, by Diana Melrose (Oxfam, 1983).

ACTIVITY: Fertabolin

1. Anabolic steroids may be used by weight lifters to increase muscle mass and strength.

3. Drugs are seen as having almost mystical curative powers. Their origin in the West gives them a special status.

4. The capital expenditure on a system to supply clean water is so high that its profitability is insignificant, compared to the high profits returned by the drug companies' investment.

---

11 ENERGY – RESOURCES AND CONSUMPTION

All available energy sources – nuclear, fossil fuel or renewable – have some kind of harmful impact on the environment. The purpose of sections 11, 12 and 13 is to highlight and compare these various impacts and to encourage students to come to their own conclusions about the relative merits of each energy source.

Energy Across the Curriculum, published by the Energy Efficiency Office, Department of Energy is an invaluable directory of resources. It gives details of over 500 items, including software, videos, tape-slide sets and written material.

The Energy Question, by Gerald Foley (Penguin, 1987) is a comprehensive and very readable account of the full range of energy resources, which takes a commendably balanced approach to controversial issues such as nuclear power.

Unit 11.1 Energy basics

This unit revises some basic principles involved in energy conversions and applies them to the problem of reducing energy losses when using fuels directly or indirectly to generate electricity.

ACTIVITY: **Energy flow diagrams**

1. a) gravitational energy ⟶ kinetic energy ⟶ electrical energy

   b) chemical energy ⟶ electrical energy ⟶ sound energy

   c) solar energy ⟶ electrical energy ⟶ electromagnetic wave energy (light)

2. a) Direct use: chemical energy ⟶ heat energy
   Indirect use: chemical energy ⟶ heat energy ⟶ kinetic energy ⟶ electrical energy ⟶ heat energy

   b) Direct use is more efficient, because there are fewer energy conversions during which heat is lost.

Unit 11.2 Fossil fuels

This unit encourages students to approach the scientists' predictions of the length of time that fossil fuel reserves will last with caution. The economic climate and political will to take action are seen as key factors in determining the future use of fossil fuels.

It also invites students to explore the safety hazards of extraction and to compare the differences in air pollution associated with the different fossil fuels. The impact of CO_2 emissions is covered in more detail in Unit 11.9.

ACTIVITY: **Reserves of fossil fuels**

1. a) 700–800 years.

   b) 110 years.

2. *A Blueprint for Survival* underestimated the quantity of fossil fuel reserves left in the Earth and overestimated the growth of future energy consumption.

3. Odell's prediction of oil consumption four or five times the 1972 level is much higher than the graph's prediction.

4. a) Higher fuel costs will lead to a greater emphasis on efficient use and therefore consumption will decrease.

   b) As fuel prices rise the probable reserves that were previously uneconomic to produce become a more certain economic proposition. They are then reclassified as proven reserves.

5. As fuel becomes scarcer and more difficult to produce the price will rise.

6. Technical problems include the difficulties in extracting coal from narrow seams and drilling for oil in inhospitable environments.

ACTIVITY: **Environmental impacts of fossil fuels**

1. The combustion of gas does not produce sulphur dioxide as coal combustion does. It avoids the need for costly flue gas desulphurisation equipment that would need to be fitted to coal-fired power stations.

2. The students will probably come up with a wide range of environmental effects. Some to consider are that extraction of fossil fuels has historically been associated with explosions and other disasters (e.g. Aberfan) that have cost the lives of both workers and local inhabitants. Long-term health risks, such as the higher incidence of lung diseases among miners, are also important. The environmental consequences of oilspills at sea and the subsequent clean up operations is a topic that provides ample opportunities for student research using the most recent such incident as a case study. Laying gas pipelines and building offshore oil and gas platforms also have environmental effects.

3. The Public Relations departments of National Power, PowerGen, the major oil companies and car manufacturers and environmental pressure groups such as Friends of the Earth have leaflets covering these issues (see 'Sources of further information').

Unit 11.3 Supplying electricity

Policy decisions in this area change rapidly. Students who wish to explore more recent statistics on the supply of electricity can contact the newly formed power generation companies National Power, PowerGen and Nuclear Electric (see 'Sources of further information'). The annual energy statistics published by the Government Statistical Office are held in reference libraries.

ACTIVITY: **Electricity supplies**

1. a) nuclear power stations — 14.4%

   b) coal-fired power stations — 75.2%

   c) oil-fired power stations — 3.5%

   d) dual (oil/coal) power stations — 7.0%

   e) gas-fired stations — 0.02%

   f) hydroelectric: natural flow — 0.11%

2. If the load factor is low it indicates that the power station is being underused. A 100% load factor is not achieved in practice because of periodic maintenance shutdowns.

3. a) Magnox   b) Gas turbine

4. Variations in load factor reflect differences in the power stations' cost effectiveness and reliability.

5. See Table 11.3.1.

6. 66%. Energy is lost in heat losses during the various energy conversions that occur in power stations (see Figure 11.1.2).

7. Thermal efficiency is only applicable when a fuel is burnt to produce heat.

Science, Technology and Society © D. Andrews, 1992

8 Scotland's mountains give more opportunity for exploiting hydroelectric power. In England and Wales the mountain ranges are lower and less extensive.

9 AGRs have the highest thermal efficiency. They are a more recent design than the Magnox reactors and have been designed to be more efficient.

Unit 11.4 International comparisons of energy consumption

This unit gives an opportunity for students to discover and analyse differences in energy consumption between different countries. Total energy consumption and differences in energy sources can be analysed. The *BP Statistical Review of World Energy*, 1986 gives data on a wider range of countries for students who wish to explore this issue in more depth.

ACTIVITY: Energy consumption

1 USA, France, Japan and the USSR.

2 USA oil consumption is over nine times the UK figure. This is only partly explained by the larger population (USA population is 238.7 million compared to the UK's 55.6 million.) It also indicates the more intensive use of private motor transport.

3 China. This is because of the relative scarcity of private motor cars and China's reliance on steam-powered technology for industry and electricity production.

4 The UK and the USSR.

5 Overall energy consumption in the UK and France is about the same. In the UK geological factors explain the greater dependence on coal and natural gas. Nuclear power is a more important source of electricity production in France than in the UK because of the relative lack of fossil fuel reserves. France's extensive mountainous regions provide scope for more hydropower.

6 Norway. This is because there are scarce coal reserves and a resistance to nuclear power. There are many suitable sites for hydropower in Norway's extensive mountain ranges.

7 These calculations will show up the wide differences in per capita energy consumption, which correlates closely with the country's level of economic development. If the efficiency of energy consumption in the developed countries improves and the standard of living in the developing countries increases, this gap will narrow.

8 Renewable energy sources are not mentioned. Countries without fossil fuel reserves are likely to make most use of these sources in the future.

Unit 11.5 Patterns of energy use in the UK

Students who wish to delve more deeply into the statistics of UK energy usage can consult the annual *Abstract of Energy Statistics* produced by the Department of Energy.

ACTIVITY

1 a) 29.4%   b) 28.3%   c) 23.0%

2 Iron and steel. Coke is consumed as a raw material in the steel production process (steel = iron + carbon).

3 Agriculture. Electricity is used to operate a wide range of farm equipment, e.g. milking parlours. Rail transport. Many lines are now electrified Miscellaneous.

4 The domestic sector. Gas is the main fuel for central heating systems.

6 See Figure 11.6.1.

7 Coal consumption is likely to drop and gas consumption to rise because of the increased emphasis on the different pollution levels they produce (see Unit 11.2). Petrol consumption in the transport sector is likely to rise, as the number of private cars increases. If energy efficiency is given active support the total energy consumption is likely to fall.

Unit 11.6 Trends in energy consumption in the UK

ACTIVITY: Energy trends (1960–1990)

1 Nuclear power and natural gas. Natural gas had only just been discovered in the North Sea and nuclear power was in the early stages of development.

2 Car ownership increased and this led to a rise in the consumption of oil.

3 a) Gas central heating replaced coal fires as the main form of domestic heating.

  b) Steam engines fuelled by coal were rapidly replaced by diesel and electric engines.

  c) Coal consumption dropped as it was no longer needed for the production of gas.

4 a) 164 − 136 = 28 mtce   b) 17%

  c) The Arab–Israeli war led to the formation of OPEC (Organisation of Petroleum-Exporting Countries), consisting of the countries that produced the bulk of the world's oil. In 1974 this consortium trebled the price of crude oil within the space of a few months.

  d) A 19% reduction in oil consumption.

5 The effect of the 1984 strike was greater than the effect of the 1974 strike.

6 In 1974 the miners' strike did not coincide with an increase in oil consumption. During the 1984 strike oil consumption increased dramatically.

7 a) The increase in oil consumption during the 1984 strike is approximately equal to the drop in the coal consumption, indicating that oil replaced coal as a fuel for power stations.

  b) Power stations that could be fuelled by either coal or oil were introduced to reduce the dependency on coal. This avoided a repetition of the politically disastrous consequences seen in 1974 when power cuts and the three-day week forced a general election.

8 a) Nuclear electricity and natural gas have increased. Oil and coal have decreased.

  b) A discussion on this question could range widely and cover the following issues: the consequences of the current freeze on the construction of nuclear power stations; the popularity of private motor transport; the prospects for improved energy efficiency in the domestic, transport and industrial sectors; the impact of policies to reduce the greenhouse effect and acid rain; the future of renewable energy technologies.

ACTIVITY: Energy trends (1977–1987)

1 Domestic and transport sectors have seen an increase in energy consumption while industry has seen a decrease.

2 Transport has seen the greatest increase in energy consumption.

3 There has been a greater emphasis on energy efficiency in industry and the use of a diverse range of domestic electrical appliances has increased.

4 More efficient use of energy and a decrease in output.

5 In industry there has been a reduction in the consumption of all energy types apart from electricity. The most significant reduction has been in oil. The growth of gas central heating is reflected in the domestic consumption graph. These trends to some extent reflect the rises in energy costs. For example, as oil prices rise, oil consumption in industry is reduced.

Unit 11.7 Energy, transport and pollution

ACTIVITY: Transport energy trends (1977–1987)

The analysis of energy use by the transport sector shows how dependent it is on oil. The long-term future of road transport in the light of scarcer and more expensive oil provides a useful starting point for a class debate or more detailed project work.

1 a) 3346 million therms   b) 715 million therms

2 Air transport.

3 Coal consumption has fallen to zero as steam engines have been phased out completely. Electricity consumption has increased by 5% and petroleum by 6.6%. This indicates that neither was particularly favoured over the other during this time.

4 The students' views will reflect their assessments of the policies that governments are likely to adopt towards road and rail transport in the future.

5 72.2%.

6 Students may wish to consider the following: 55 mph (90 km/h) speed limit on motorways; tolls on cars entering large cities; compulsory fitting of catalytic converters; banning cars from city centres.

ACTIVITY: The switch to unleaded petrol

2 a) Unleaded: 6.6%, Leaded: 93.4%
  b) Unleaded: 30.3%, Leaded: 69.7%
  c) Unleaded: 38.6%, Leaded: 61.4%

3 The price differential increased over this period, reaching about 2p per litre in 1991 as a result of tax changes in the Budget. Public awareness of the dangers of lead increased, as a result of publicity.

4 The lower consumption in the winter months reflects the reluctance to travel on long car journeys at this time of year. There is also less holiday traffic.

5 a) 4th quarter 1991   b) 1st quarter 1995

6 The fact that some older cars are not suitable for conversion to unleaded petrol will prevent the 100% level of sales being reached by the predicted date. The Department of Energy can provide the latest figures on petrol sales (see 'Sources of further information').

8 Students may like to suggest the optimal price differential to discourage sales of leaded petrol.

9 Suggestions may include: legislation, increased price differential.

Unit 11.8 Save it!

This unit emphasises the financial incentives for saving energy. The important environmental consequences of improving energy efficiency are dealt with in Unit 11.9.

The Energy Efficiency Office at the Department of Energy publishes a number of free leaflets that give detailed information on a wide range of energy-saving measures for homes, as do environmental campaigning groups, such as Friends of the Earth (see 'Sources of further information').

Unit 11.9 Global warming

The underlying question in this unit is whether the population of the world's affluent countries will accept a change in living standards to safeguard the future of the planet. This provides a suitable topic for a class debate.

Blueprint for a Green Planet, by J Seymour and H Girardet (Dorling Kindersley, 1987) is a very readable, practical guide for students who wish to explore further the positive steps required to conserve the planet's resources.

ACTIVITY: Decisions on global warming

2 Practicability and public acceptability are likely to be the main issues for students. Some consideration of the economic feasibility and impact of each proposal could also be attempted.

4 The impact of Third World economic growth on the greenhouse effect could be considered here. For example, the outcry that would occur if limiting car ownership in the developing countries was seen as a solution to the greenhouse effect, particularly as this effect has been largely caused by the lack of foresight of the developed countries.

12   RENEWABLE ENERGY RESOURCES

Review, a quarterly journal of renewable energy produced by the Department of Energy, is a valuable source for the latest news of research projects, commercial developments and government policy. See 'Sources of further information'.

12.1   Solar energy

ACTIVITY: Passive solar design

1 Layout (b).

2 A flat glass skylight allows light to enter directly and can cause problems of glare and overheating. Light entering through a clerestory is diffused by baffles and these problems are avoided.

3 Solar gain is the exploitation of solar radiation to replace other energy sources for heating and lighting.

4 a) A conservatory provides a large surface area of glass, allowing optimum solar heating of the air it encloses.

   b) Large south-facing windows maximise the solar gain as the sunlight can enter directly throughout the day.

c) The north side of the house is always in shade and therefore cooler. Heat loss from the house is reduced by having smaller windows on this side.

5 a) On day 1 solar heat gain provides a relatively small proportion of the house's heating requirements compared to the auxiliary heating. On day 2 the situation is reversed. This would be because day 2 was sunnier than day 1.

   b) The overall contribution of auxiliary heating is less on day 3 compared to day 2. The contribution of solar gain is about the same on the two days. This is explained by the rise in the external air temperature over the two days.

6 a) Incidental gains from cooking, lighting and the body heat of the inhabitants.

   b) These activities are not influenced by seasonal variation.

7 May to October.

8 a) 29%   b) 40%

9 a) Solar gain can reduce the need for auxiliary heating even on cold days.

   b) In the winter months, when the heating demands are highest, useful solar gain is lowest.

Unit 12.2   Biomass energy

ACTIVITY: Using biomass energy

Visits to sewage treatment plants can provide useful first hand experience of sewage digesters.

1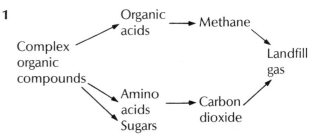

2 The relatively low calorific value of landfill gas means the efficiency of the energy conversions in indirect electricity generation is too low for commercial exploitation.

3 Incineration releases many air pollutants, such as the carcinogenic polycyclic hydrocarbons. Landfill gas burns relatively cleanly. Incineration reduces the volume of accumulated waste; landfill gas does not.

4 Cheaper than oil and is not dependent on a non-renewable resource.

5 Information on small-scale biomass energy can be obtained from the Intermediate Technology Development Group (see 'Sources of further information').

Unit 12.3 Hot rock power: geothermal energy

ACTIVITY: **The potential of hot rock power**

1. Granite often contains radioactive elements, which give off heat as they decay.

2. In many areas the temperature gradient in the Earth's surface is not steep enough for exploitation of geothermal power to be economically feasible.

3. Volcanoes cause very high underground temperatures that are a source of direct and indirect steam production. This is considerably cheaper and less polluting than burning fossil fuels to generate heat.

Unit 12.4 Tidal power: a feasibility study

ACTIVITY: **The Severn Barrage**

1. a) A tidal power station would cut carbon dioxide emissions by replacing several large coal-fired power stations, thus reducing the build-up of greenhouse gases.

   b) The amount of sediment on the bed of the Severn Estuary on the landward side would build up, as its route to the sea would be blocked by the barrage.

   c) The reduced strength of the currents in the basin created by the barrage would allow the sediments to settle. The less turbid water would improve the depth of light penetration and allow the algae population to increase. Predictions suggest that this would have a 'knock on' effect on the invertebrate population and therefore increase the availability of food for birds and fish.

   d) If the study's prediction of a local economic boom is fulfilled, more jobs will become available in recreation and leisure and unemployment will fall.

   e) Land and property values may increase in line with the predicted economic regeneration of the region (but see question 2b).

2. a) Favourable: the development of leisure and recreation activities would attract tourists.

   b) Opposed: the increased activity and noise would detract from the attractiveness and the value of the house.

   c) Favourable: improvement in journey time and petrol cost reduction.

   d) Favourable: the reduced tidal range in the basin would increase the opportunities for sailing.

   e) Favourable: an economic boom would increase demands for building and construction work.

Unit 12.5 Hydropower: large-scale and small-scale

ACTIVITY: **Hydropower problems**

1. Less pollution, no fuel costs and useful potential for irrigation and flood control.

2. The build-up of sediment in the turbines eventually causes damage that is uneconomic to remedy.

Unit 12.6 The potential of wind power

ACTIVITY: **The impact of wind power**

1. Load factor

$$= \frac{\text{Annual electricity supplied (GW h)} \times 100\%}{\text{Hours in the year } (365 \times 24) \times \text{DNC}}$$

$$= \frac{9 \times 100}{365 \times 24 \times 3}$$

$$= 3.4\%$$

2. The Burgar Hill 3 MW wind turbine produces 9 GWh annually. To meet 10% of the UK's energy demand, 25 000/9 or about 2780 similar capacity turbines would be required.

3. $2780/5 = 560 \text{ km}^2$ correct to 2 significant figures.

4. a) and b) The environmental impact of wind, nuclear and fossil fuel power stations will be the major underlying issues.

5. Remote unspoilt regions attract tourists seeking natural beauty. Noisy and unsightly wind farms would provide an unwelcome reminder of the environment that the tourists are trying to escape from.

6. The difficulties of constructing and maintaining a turbine in sometimes treacherous sea conditions would lead to high costs compared to land-based turbines.

Unit 12.7 Wind power in the developing world

Technical reports on the use of a wide range of alternative and renewable technologies in developing countries are available from the Intermediate Technology Group (see 'Sources of further information').

ACTIVITY: **Wind power feasibility study**

1. a) 0.12 kW h
   b) 0.41 kW h
   c) 0.97 kW h

2. A doubling in wind speed produces an eight-fold increase in energy output.

3. $(6 \times 0.25) + 0.5 + 1.0 = 3.0$ kW h

Science, Technology and Society © D. Andrews, 1992

4 Output of one generator
   = 0.0048 × 1 × 3.14 × 3³ kW h
   = 0.41 kW h

   Number of generators required

   $= \dfrac{\text{Total required output}}{\text{Ouput of a single generator}}$

   = 3.0/0.41
   = 7.3

   Eight generators would be required.

5 a) 3.0 = 0.0048 × a × 27
      a = 23.15 m²
   b) 23.15 = 3.14 × r²
      r = 8.5 m
      Blade diameter = 17 m

6 a) Wind generator.
   b) Diesel generator.
   c) With high mean wind speeds, wind power is cheaper than a diesel-powered generator. At lower wind speeds, wind power is only cost effective if the required energy output is low.

Unit 12.8 Wind and wave power under attack

ACTIVITY: **Views on wind and wave power**

1 a) Tonga is an island in the Pacific Ocean.
   b) It has strong currents and a large choice of sites for development.

2 a) The wave power technology developed by the Norwegian company supplying the Tonga scheme was being questioned after a storm had destroyed a similar wave power station.

3 Neither produce carbon dioxide, a major greenhouse gas.

4 a) Their enthusiasm for alternative energy blinds them to any criticisms against it.
   b) The editorial suggests that studies showing that alternative energy technology is uneconomic are regarded by supporters of AT as 'fixed', i.e. politically motivated.

5 a) The potential loss of life is much greater in nuclear reactor and oil rig accidents than in wave or wind power station accidents.
   b) It supports the view that renewable energy technology is safer.

6 a) Load factor measures the ratio of actual power generation to the maximum that could potentially be achieved.
   b) A load factor of 9.1% is much lower than the load factors for fossil fuel or nuclear power stations.
   c) Wind speed is highly variable.

7 The researchers would now realise that a stronger construction was needed.

8 There is little financial incentive to make energy savings when the cost of oil is low.

9 Economic and technical factors such as difficulties of erecting offshore platforms and maintaining them; increased costs of this work.

Unit 12.9 The future of renewable energy technologies

ACTIVITY: **Renewable energy policy**

1 These technologies are not suited to electricity generation. However they are good sources of heat, which can be used directly.

2 See Units 12.1 to 12.8.

3 One.

4 24%

ACTIVITY: **Economics of renewable energy**

1 High fossil fuel costs will make it economically more attractive to invest in renewable energy technologies.

2 Subsidies can be used to promote technologies that are in the early stages of development and have beneficial environmental effects.

13 NUCLEAR POWER

Unit 13.1 What is radioactivity?

Multiple Exposures – Chronicles of the Radiation Age, by Catherine Caulfield (Penguin Books, 1990) is a well-balanced account and an excellent source of case studies of human exposure to radioactivity.

ACTIVITY

1 a) 9000 million years
   b) 60 years
   c) 110 seconds

2 a) Uranium ore contains mainly uranium-238. This only emits alpha particles, which cannot penetrate the skin.
   b) Uranium dust enters the lungs where alpha particles cause cancer.

3 Carbon-14. With a half life of only 5730 years, carbon-14 would not be present in a measurable quantity after this time.

4 As iodine-131 has a half life of 8 days, after 4 weeks the quantity of it would have diminished to almost $\frac{1}{16}$ its original level and would not pose a health threat.

Unit 13.2 How do nuclear power stations work?

Nuclear Electric, Nuclear Electricity Information Group, Greenpeace and Friends of the Earth can provide useful material to extend this topic (see 'Sources of further information').

Unit 13.3 Reprocessing and breeder reactors

ACTIVITY

1 The Calder Hall reactor produced plutonium for nuclear warheads.

2 Reprocessing cuts the amount of waste by retrieving fission products that can be used in breeder reactors. This has to be weighed against the threat of terrorists gaining access to the plutonium that has been separated.

3 BNF have produced leaflets and a video that seek to reassure the public of the safety of nuclear flasks. A less optimistic view is given in literature produced by groups such as the London Nuclear Information Unit (see 'Sources of further information').

Unit 13.4 What can we do with radioactive waste?

ACTIVITY: Options for radioactive waste disposal

Space Advantages: if successful it avoids contaminating the Earth's surface. Disadvantages: a rocket failure after launch would spread radioactive waste over a wide area.

Ocean dumping Advantages: low cost. Disadvantages: risk of pollution if the steel corrodes.

Burial 500 m below the land surface Advantages: avoids the problems of surface storage. Disadvantages: risk of leakage leading to contamination of groundwater.

Burial 500 m below the sea bed Advantages: leakage would not enter the groundwater used for drinking. Disadvantages: high cost of tunnelling.

Store at guarded sites on the surface Advantages: no uncertainty about leakage. Disadvantages: risk of contamination due to accidents or terrorism.

ACTIVITY: The radioactive waste problem

1 UKAEA. By excluding intermediate and low-level radioactive wastes and expressing the volume as a per capita figure.

Unit 13.5 The Chernobyl accident

It is advisable to provide a wide range of sources to build up a balanced picture of the safety of nuclear power stations. For addresses, see 'Sources of further information'.

A more detailed analysis of the background to Chernobyl can be found in *The Chernobyl Disaster*, by V Haynes and M Bojcun (The Hogarth Press, 1988). It gives a balanced view of the contributions of the political, human and engineering factors to the accident.

Power production: what are the risks? by J H Fremlin (Oxford University Press, 1987) gives a detailed comparison of the different methods of power production and also includes technical details of the cause of the Chernobyl accident. The author sets out to show that the dangers of nuclear power have been exaggerated.

ACTIVITY

1 The first four quotes confuse the radioactive isotopes released by the explosion with the ionising radiation they emit.

The other quote shows misunderstanding of the prophylactic action of potassium iodate tablets against the take-up of iodine-131.

3 Estimation is hampered because of differences in quantifying several key variables: the total quantity of radioactive material released; the range of radiation doses received by the populations of European countries; the impact of a particular dose on an individual's risk of suffering a fatal cancer.

4 Students' responses to this question will depend on whether they see the 1000 deaths as insignificant against the total of 15 million.

Unit 13.6 Nuclear power and leukaemia: problems of extrapolation

A more detailed treatment of the impact of radiation on health is given in *Radiation Risks: An Evaluation*, by David Sumner (Tarragon Press, 1987). Interested students could study the report of the government's advisory group on the leukaemia cluster around Sellafield, *Investigation of the Possible Increased Incidence of Cancer in West Cumbria* (HMSO, 1984). This unit is closely linked to Unit 4.3, Medical effects of nuclear war.

ACTIVITY: Health risks of radiation

2 Linear.

3 Nuclear power workers receive lower doses over decades, whereas the Nagasaki and Hiroshima survivors received a very high dose over a very short period.

4 The linear line gives the highest risk.

Unit 13.7 Hot and cold fusion power

ACTIVITY: Energy from fusion

1 In a fission reaction a single atom splits apart. In a fusion reaction two atoms fuse together.

2 The plasma must be kept at a particular density at a temperature of 100 000 000°C for at least one second.

3 In the long term fusion has the politically attractive promise of limitless power with little environmental pollution. Governments are prepared to make a financial gamble because the potential pay-off is so high.

4 Pooling financial resources and scientific expertise is essential on a project like this, where the costs are so high and the technical difficulties so complex.

Sources of further information

The Advertising Standards Authority
Brook House
Torrington Place
London WC1

The Association of the British Pharmaceutical Industry
12 Whitehall
London SW1A 2DY

BP Educational Service
BP Co Ltd
Britannic House
Moor Lane
London EC2 9BU

British Coal
Public Relations Department
Hobart House
Grosvenor Place
London SW1X 7AE

British Gas plc
Head Office
Rivermill House
152 Grosvenor Road
London SW1

British Nuclear Fuels
Risley
Warrington WA3 6AS

Brook Advisory Centres
153A East Street
London SE17

Building Research Advisory Unit
Building Research Establishment
Garston
Watford WD2 7JR

Council for Environmental Education
School of Education
University of Reading
London Road
Reading RG1 5AQ

Cystic Fibrosis Research Trust
Alexandra House
5 Blyth Road
Bromley
Kent BR1 3RS

Data Protection Registrar
Springfield House
Water Lane
Wilmslow
Cheshire SK9 5AX

Department of Energy
Thames House South
Millbank
London SW1P 4QJ

Down's Syndrome Association
153-5 Mitcham Road
London SW17 9PG

The Education Secretariat
Institute of Energy
18 Devonshire Street
London W1N 2AU

The Electricity Council
30 Millbank
London SW1P 4RD

Energy Efficiency Office
Department of Energy
Thames House South
Millbank
London SW1P 4QJ

Family Planning Association
27-35 Mortimer Street
London WC1N 7RJ

Friends of the Earth
26-28 Underwood Street
London N1 7JQ

Greenpeace
Canonbury Villas
London N1 2PN

Health Education Authority
Hamilton House
Mabledon Place
London WC1H 9TX

Intermediate Technology Development Group
Myson House
Railway Terrace
Rugby CV21 3HT

International Planned Parenthood Federation
Regent's College
Regent's Park
London NW1 4NS

Leukaemia Research Fund
43 Great Ormond Street
London WC1 3JJ

London Nuclear Information Unit
Camden Town Hall
141 Euston Road
London NW1 2RU

Science, Technology and Society © D. Andrews, 1992

Marlow Foods Ltd
9 Station Road
Marlow
Bucks SL7 1NG

MENCAP
123 Golden Lane
London EC1

National Anti-Vivisection Society
51 Harley Street
London W1

National Centre for Alternative Technology
Llwyngwern Quarry
Machynlleth
Powys SY20 9AZ

National Power
Department of Information and Public Affairs
Sudbury House
15 Newgate Street
London EC1A 7AU

National Radiological Protection Board
Chilton
Didcot
Oxon OX11 0RQ

Nuclear Electric
Bedminster Down
Bridgwater Road
Bristol
Avon BS13 8AN

Nuclear Electricity Information Group
22 Buckingham Gate
London SW1

Organisation for Sickle Cell Anaemia Research (OSCAR)
109 Mayes Road
Cambridge House
Wood Green
London N22

Oxfam
274 Banbury Road
Oxford OX2 7DZ

Parkinson's Disease Society
22 Upper Woburn Place
London WC1H 0RA

Population Concern
231 Tottenham Court Road
London W1P 9AE

PowerGen
Haslucks Green Road
Solihull
West Midlands B90 4PD

RSPCA
Causeway
Horsham
West Sussex RH12 1HG

Renewable Technology Enquiries Bureau
Energy Technology Support Unit
Building 156
Harwell Laboratory
Oxon OX11 0RA

Review
Editorial Office
Room 3.2.21
1 Palace Street
London SW1E 5HE

Shell Education Service
Shell-Mex House
Strand
London WC2R 0DX

Society for the Protection of Unborn Children
7 Tufton Street
London SW1P 3QN

The Soil Association
86 Colston Street
Bristol BS1 5BB

The Spastics Society
12 Park Crescent
London W1

TACADE
3rd Floor
Furness House
Trafford Road
Salford M5 2XJ

Terrence Higgins Trust
BM/AIDS
London WC1N 3XX

Transport 2000
Walkden House
10 Melton Street
London NW1 2EJ

United Kingdom Atomic Energy Authority
Information Services Branch
11 Charles II Street
London SW1Y 4QP

United Kingdom NIREX Ltd
Curie Avenue
Harwell
Didcot
Oxon OX11 0RH